W9-ASP-706

WITHDRAWN

Illinois Central College
Learning Resource Center

THE SCIENCE OF HUMAN COMMUNICATION

THE SCIENCE OF HUMAN COMMUNI- CATION

New Directions and New Findings in
Communication Research

Edited by WILBUR ₍Lang₎ SCHRAMM, 1907 −

BASIC BOOKS, INC., PUBLISHERS
New York   *London*

Fourth Printing

© 1963 by BASIC BOOKS, Inc., Publishers
Library of Congress Catalog Card Number 63–12849
Manufactured in the United States of America
Designed by Michel Goldberg

Editor's Foreword

The Voice of America from time to time broadcasts a series of talks by American scholars, intended to sum up authoritatively for foreign listeners some of the latest scholarly insights and research findings on a topic of wide interest. In 1961, I was asked by officials of VOA to serve as chairman of such a series on communication research. The resulting talks were broadcast on the Worldwide English service of VOA in the spring of 1962 and were several times repeated. They aroused broad interest and resulted in a considerable amount of correspondence with listeners. Now Basic Books has gathered all these talks into one volume, so that they can be more readily available.

In planning the series, we made no attempt to prepare a systematic textbook on human communication or to represent all the areas, problems, and findings of communication research. Rather, we selected certain areas where new findings were available which we thought would be widely interesting to listeners and readers, and where outstanding scholars were available to talk about their specialties. Of course, these authors have written at greater length and in more detail about their respective subjects in more technical publications. The most relevant of these writings are listed at the end of each writer's discussion. A few scholars we had hoped to present to VOA's overseas audience were unable to participate. Among these was one of the founders of communication research, Professor Carl Hovland of Yale, who had hoped to take part but became ill and died during the time the series was being planned, and to whose memory and great career this book would properly be dedicated.

These talks are therefore an introduction to the problems, the findings, and some of the scholars in research on human communication. Obviously, they do not represent *all* the problems, the findings, or the scholars, but each talk is followed by a few suggestions for reading which will help to fill in the gaps.

The great courtesy and helpfulness of the Voice of America personnel, and particularly of Mr. Walter Nichols, director of the Forum Series of which these talks were a part, must be gratefully acknowledged.

WILBUR SCHRAMM

Contents

THE SCIENCE OF HUMAN COMMUNICATION

1 COMMUNICATION RESEARCH IN THE UNITED STATES

Wilbur Schramm

What do we mean by "communication research"? How long has it been in existence? Who started it? What problems does it deal with? And, in broad terms, what has research so far told us about how communication works and produces effects?

These are the kinds of questions taken up in this introductory paper by Wilbur Schramm. The author was born in 1907 in Marietta, Ohio. He earned degrees from Marietta College and Harvard and his Ph.D. in 1932 from the University of Iowa. He has taught at Iowa, at the University of Illinois, where he was founder and director of the Institute of Communications Research and Dean of the Division of Communication, and since 1955 at Stanford, where he is Janet M. Peck Professor of International Communication and Director of the Institute for Communication Research. Among his publications have been Mass Communication *(1949, 1960),* Process and Effects of Mass Communication *(1954),* Responsibility in Mass Communication *(1957),* One Day in the World's Press *(1959), and* Television in the Lives of Our Children *(1961). He has engaged in research or consultative studies on four continents and in many countries.*

During the last thirty years an increasing number of scholars in the United States have become interested in studying the process and effects of communication. Communication, of course, has not become an academic discipline, like physics or economics, but it has become an extraordinarily lively area of research and theory. It is one of the busiest crossroads in the study of human behavior, which is understandable because communication is a—perhaps *the*—fundamental social process. Without communication, human groups and societies would not exist. One can hardly make theory or design research in any field of human behavior without making some assumptions about human communication.

Communication theory and research have therefore attracted the interest of psychologists, sociologists, anthropologists, political scientists, economists, mathematicians, historians, and linguists, and men from all these fields and more have made contributions to our understanding of it. It has been an academic crossroad where many have passed, but few have tarried. Psychologists have worked in the problems of communication for a while and then gone back to problems which are more distinctively psychological. Mathematicians developed information theory and then turned to theory and problems more distinctively mathematical. And so with the other disciplines: communication has been an auxiliary study, necessary to the understanding of human and social behavior, and contributory to other theory. Nevertheless, out of the comings and goings at this academic crossroad, and from the comparatively small number of scholars who have devoted their whole careers and energies to the study of human communication, has come an impressive shelf of books and articles, some of which will be reflected in the papers that follow.

Four men have usually been considered the "founding fathers" of communication research in the United States. Two of these were psychologists, one a sociologist, one a political scientist. Two of them were European-born and -educated, but came to this country very early in their careers.

One of the latter was Paul Lazarsfeld, a sociologist trained in Vienna, who came to the United States in 1932 and became deeply interested in the audiences and effects of the new media of mass communication. When he entered mass media research, the broadcasters and advertisers of this country had already begun to measure audiences in order to find out how well the radio was doing—how many listeners it was attracting, and how well they liked what they heard.

To Lazarsfeld, it became immediately apparent that it was just as easy and far more significant to use audience measurements to study the audience as to study the medium.

That is, the programs people select tell us something about the people, as well as about the programs. The next step was to find out *why* they choose what they choose to listen to. And then to find out how they use what they get from the mass media, and what effect the media have on their voting habits, their tastes, and their general orientation toward life and society.

This is the line of research that Lazarsfeld pursued—audience studies, voting studies, studies of campaigns, studies of mass media effect, and studies of personal influence as related to mass media influence. He founded one of the most influential survey research organizations in this country—the Bureau of Applied Social Research at Columbia University —and for more than twenty years that Bureau has continued to produce studies of high quality and to train outstanding young scholars. Two of Lazarsfeld's students, graduates of the Bureau who are now outstanding scholars in their own right—Dr. Elihu Katz and Dr. Joseph T. Klapper—have contributed lectures to this series, and Dr. Lazarsfeld himself, in collaboration with his colleague, Dr. Herbert Menzel—another of his former students—is the author of one of the lectures.

The other European-born member of the "founding fathers" was Kurt Lewin, the Gestalt psychologist, also trained in Vienna, also an immigrant into this country in the early 1930's, who exerted a great influence on students at the University of Iowa and at the Massachusetts Institute of Technology. Lewin's central interests were communication in groups and the effect of group pressures, group norms, and group roles on the behavior and attitudes of their members. He was an extraordinarily ingenious experimenter and had the ability to attract imaginative and brilliant students. The Group Dynamics movement, if it can be called a movement, in the United States is the shadow of Lewin. The nature of his influence on communication research can be judged from the contribution to this series made by Dr. Leon Festinger,

3

one of his students, who has developed the theory of cognitive dissonance. Lewin himself died at an untimely age, but his influence lives on.

Lazarsfeld was a sociologist, specializing in survey research, interested in the effects of the mass media and their relation to personal influence. Lewin was a psychologist, specializing in experiments, interested in the workings of human groups.

A third member of the "founding fathers" was Harold Lasswell, a political scientist, trained at the University of Chicago, and for many years a teacher there and at Yale. He was neither a survey researcher nor an experimenter; rather, his method was analytic. He pioneered in the study of propaganda, in the large systemic examinations of communication in nations and societies, and in the study of influential political communicators. But perhaps he will be remembered longest in this field because of his development of scientific content analysis. Dr. Lasswell was unable, regrettably, to appear in this series, but it includes a contribution by one of his pupils, Dr. Ithiel de Sola Pool, of M.I.T.

The fourth member of this group was Carl Hovland, who was trained at Yale as a psychologist. Before World War II, he had already made a reputation as an experimental psychologist. Called into the Army research program in 1942, he became deeply interested in communication and attitude change. When the war was over, he abruptly changed the direction of his career, returned to Yale, and organized a research program on communication and attitude change. His colleagues and students in the program were some of the best young psychologists in the field.

Hovland's method was a tight, careful, experimental one, varying a single element at a time, controlling the others, testing hypothesis after hypothesis, building up slowly but surely a systematic theory of communication. What he was doing, in effect, was building a modern scientific rhetoric. Many of the problems he studied were as old as the *Rhetoric* of Aristotle. He was studying, for example, the effect of having a credible or prestigious communicator, a one-sided or a two-sided message, strong fear appeals vs. weak fear appeals, methods of "inoculating" people against propaganda, and so forth.

The books that came out of this Yale research program between 1950 and 1961 represent the largest single contribution to communication theory any man has made. Let me recall the titles of some of these books: *Experiments on Mass Communication, Communication and Persuasion, The Order of Presentation, Personality and Persuasibility, Attitude Organization and Change.*

Hovland died in 1961 of cancer, at the age of forty-eight, but his colleagues and students are carrying on the tradition of his communication research. Dr. Irving L. Janis, who succeeded Hovland as director of the Yale program, has contributed to this series, and Dr. Nathan Maccoby, who was a colleague of Hovland's in the Army research program, discusses in this series the subject which occupied Hovland all his later life—the theory of how to change attitudes by means of communication.

These four strands of influence are still visible in communication research in the United States, but increasingly they have tended to merge. Young researchers in the field now tend to be eclectic. They combine the interests of Hovland and Lewin, or the method of Lazarsfeld with the interests of Lasswell, or form some other combination. But several characteristics of communication research in this country will be evident to any persons who come from another tradition. For one thing, communication research in the United States is quantitative, rather than speculative. Its practitioners are deeply interested in theory, but in theory they can test—and they want to test it. Thus, essentially, they are behavioral researchers: they are trying to find out something about why humans behave as they do, and how communication can make it possible for them to live together more happily and productively. It is therefore not surprising that a number of our communication researchers lately have turned to the problem of how the nations of the world can communicate efficiently, and how communication can help them understand each other and live in peace.

Another characteristic of the growth of communication research in the United States has been the appearance of research centers. We have already mentioned the Bureau of Applied Social Research at Columbia, and the Communication and Attitude Change Program at Yale. In addition to these, there are now research centers of considerable size and productivity at Stanford, Illinois, M.I.T., Michigan State, and Wisconsin, and smaller ones at several other universities.

Let us turn now from this brief overview of the development and state of communication research in the United States to the substance of the research itself. In this first paper, it seems appropriate to say something about just what we mean by the term communication, and how the communication process works.

Let me hasten to say that we are not talking merely about mass communication as typified by newspapers, television, and so forth. In the United States, communication research is concerned with *all the ways* in which information and ideas are exchanged and shared. Thus we are talking about both mass and interpersonal communication. We are talking about the spoken word, signal, gesture, picture, visual display, print, broadcast, film—all the signs and symbols by which humans try to convey meaning and value one to another.

The process is the same whether the signs are broadcast on a television wave or whispered by a young man into his sweetheart's ear. The mass medium is merely a communicator in which the ratio of output to input is very large. In some individuals—teachers, public speakers, gossips, and rumor-mongers, for example—the ratio of communication output to input is also relatively large; but even compared to these individuals, the output–input ratio of a mass medium is enormous. Furthermore, the mass medium is a communicating organization—a working group of people trained and organized so as to speak with a single voice and to display a kind of corporate personality. But, except for being more complicated, what happens in the case of the mass medium is precisely parallel to what happens in a communicating individual. They both select and decode messages, encode and transmit messages, and elicit responses.

In its simplest form, the communication process consists of a sender, a message, and a receiver. The sender and the receiver may even be the same person, as happens when an individual thinks, or talks to himself. But the message is at some stage in the process separate from either sender or receiver. There comes a time when whatever we communicate is merely a sign that stands for some meaning to the sender, and that stands to the receiver for whatever meaning he reads into it. That is, at one stage in the communication process the message is merely ink on paper (as in a printed book), or a series of condensations and rarefactions in the air (as in the spoken word), or reflected light waves (as in communication by picture).

These signs have only such meaning as, by agreement and experience, we give them. For example, a printed word in a language we do not know may have little or no meaning for us. A glance from a wife to a husband may have a secret meaning that only those two people can know. On the other hand, a red traffic light will probably have the same meaning to all automobile drivers, and a scream of terror will probably have about the same meaning anywhere.

This is one of the basic principles of general communication theory: that signs can have only such meaning as an individual's experience permits him to read into them. We can make a message only out of signs we know, and we can give those signs only such meaning as we have learned for them. We can decode a message only in terms of the signs we know and the meanings we have learned for them. We call this collection of experiences and meanings a "frame of reference," and we say that a person can communicate only in terms of his own frame of reference.

For example, if a primitive man had never seen or heard of an airplane, and one were suddenly to appear, he would have to interpret it only in terms of experiences he had had with flying things. He would probably think of it as a great bird, and the pilot perhaps as a supernatural figure able to tame gigantic birds. An American teacher in Africa told me that the first few times she called the roll of names in class, the students laughed, and she worked desperately hard to learn to pronounce the names correctly, because she thought they were laughing at her unskillful way of saying their names. But when she became very skillful, they laughed all the harder. And finally she learned that they were laughing out of friendliness and pleasure at how hard she was trying and how well she was doing. In *her* frame of reference, a laugh at that point meant derision; in *their* frame of reference it meant something else entirely.

If we recall how different are the experiences of different kinds of people, and especially how different the frames of reference are between countries that are far apart and have different values and cultures, we can readily understand why it is hard to communicate between such senders and receivers, and why misunderstanding often occurs.

Simple as it may seem, a message is a very complicated thing. Not only do its signs have different meanings for different people; they also have two different kinds of meaning. One of these is *denotative* meaning: the common or dictionary meaning, which will be roughly the same for all people who use the same dictionary or go to the same school. Another is *connotative*: the emotional or evaluative meaning—how good, how powerful, how active, how dangerous something is. This varies greatly with individuals, and even with time. Obviously a hammer and sickle will have a different connotation, although perhaps the same denotation, to a Communist and a non-Communist.

Furthermore, a message has both a surface meaning and a latent meaning. When we say "Good morning," we usually do not mean anything about the blueness of the morning sky or the brightness of the morning sun; rather, we are saying something about our social relationship to the receiver. We are saying, "We are still friends," or, "I am glad to see you," or something like that. Many messages take their important meaning from the context of the relationship of the sender and the receiver, and that is why it is sometimes dangerous to interpret what is said in terms of what the words "mean," without considering the latent meaning.

Another characteristic of a message is that it is usually a number of parallel messages. For example, if you were to hear me speak, you would be hearing not only bits of language called words; you would also be hearing the intonations or pitch patterns I gave the sentences and you would notice what words I said loudly for emphasis, and where I paused. From my accent, you would get a message about where I grew up. From the quality of my voice, you would get some impression of me. If you were to see me on television, you would get still other parallel messages from my gestures, the clothes I wore, the fact that I smiled or frowned, looked serious or amused.

You may say that is all very well if a person is talking to you, but do you get parallel messages like that from, say, a printed page? Yes, you do. You see patterns of ink which you interpret as words. But these patterns have characteristics of their own. Different type faces have connotations: some seem graceful, others strong, some rough and blunt, others brisk and businesslike. The size of the type means something about the importance of what you are reading. The quality of the paper tells you something. The amount of white space makes a difference in the way you interpret the print. If there is a picture along with the text, that carries a separate message. If there is a headline or title, this is one of the ways by which almost all communication is "indexed" for us, and by which we are given a preview of its meaning.

The point I am making is simply that the impact of any message depends on more than any one single channel, actually on many channels and cues that we hear or see simultaneously. And with every message comes an especially important cue—the knowledge of *who* said it, which helps us to determine whether to accept it and act on it.

Now let us consider what happens when a message is transmitted. Let us suppose that the message has been encoded and sent and exists in the form of ink on paper, or waves in the air, or some other signs in which the sender has tried to communicate certain meanings. It must be pointed out that each of us is surrounded by many more such messages than he can possibly accept. The ratio is at least hundreds to one; it may be millions to one. Therefore, the first question is, will a receiver pay attention to the message? Will he hear the voice, or select the news story to read, or tune his radio set to a certain station at a certain time? This is determined by how readily available the message is, and by what rewards it promises. A music lover might drive fifty miles to hear a great symphony orchestra, rather than watch football at home on his television set. On the other hand, a football fan, given the choice of a game fifty miles away and the same game at home on his television, would probably take the game on television.

So the first obstacle the message must surmount is being selected out of all the competing messages. If it passes that hurdle, it will then either be accepted into or rejected from the cognitive part of the receiver. This will depend on how it is interpreted. We have already said that a message can be interpreted only in terms of the stored experience—the frame of reference—of the receiver. But we have not said enough about how an individual stores the experiences of his life. From the earliest time that he realizes his own individuality, he stores up things he believes and things he values, and these become increasingly precious and important to him. He will go to almost any length to defend these ego-related beliefs and values. For example, he will reject a message. He will, unwittingly, misinterpret a message. He will distort it. One remarkable piece of research (the "Mr. Biggott" studies) showed that people who were strongly prejudiced on a certain subject would interpret an attack on their prejudice exactly opposite to the way it was intended; they interpreted it as praise and support for their viewpoints. So the question of acceptance or rejection depends on how the message fits into the sender's values and beliefs, and how effective it is in making a place for itself among them. Some of this process is rational and some is below the level of conscious thought.

But there is still another hurdle the message must jump, if it is going to accomplish much with an individual. It must face the test of his group norms and beliefs. Just as an individual stores up beliefs and values that become a part of his personality and tries to defend them at all costs, so does he come to value group memberships—his family, his peer group, his work group, organizations he would like to belong to, and so forth. Almost any important message he receives will be relevant to one of these valued groups. We call this the "reference group" for this particular topic. If he values this reference group, he will try to match up the message with what that group believes and values. If it is in disagreement with the group norms, then it is very unlikely to be accepted without substantial change.

It makes a great deal of difference, therefore, whether a message comes into old and strongly defended territory, whether it deals with a topic on which the receiver and his reference groups hold strong positions or a topic on which there has yet been no occasion to take a firm stand. In a new area, weakly defended, there is a good chance that a well-made message will accomplish what it is intended to. In an area where the defenses are high, it will probably accomplish most, not by trying to storm the walls, but by chipping off a brick or two. We call this "canalizing" attitudes, by which we mean that it is often more effective to take existing attitudes and try to *redirect* them slightly than it is to meet them head on. This is what happened in the famous incident in which Kate Smith sold so many millions of dollars worth of war bonds by means of radio appearances. The planners of these broadcasts took advantage of attitudes toward the war effort, thrift, and civilian sacrifice, attitudes that were strongly held and fundamentally favorable to Miss Smith, and "canalized" these toward buying bonds. On the other hand, when the Orson Welles radio play about a fictional invasion from Mars caused panic, the people who panicked were those who had no effective defenses. For them, it was a real experience; they trusted radio "news reports" implicitly; they had no built-in responses ready for an invasion from Mars and were unaccustomed to check broadcasts against other evidence. So when the broadcast entered this relatively new and unde-fended area, they accepted what they heard and ran for the hills!

In a new area, therefore, a well-made communication from a trusted source can accomplish considerable change. In an old, well-defended area, it is very unlikely that a communi-cation will accomplish sharp and abrupt change unless it is accompanied by equally significant changes in the surround-ing situation. For example, a soldier who will pay little atten-tion to a surrender leaflet when his army is intact may read it carefully when his unit is beaten and broken up. And many a convert has been helped to a radical change in belief by being welcomed into a new group with radically different ideas. Each of these constitutes a change in situation that makes different attitudes and opinions more easily tenable.

But attitude change through communication is a very complex subject, and several other papers in this series will deal with it. Therefore, let me say merely that in order to ac-complish anything of any considerable importance with a message, a sender must get it chosen and attended to by the receiver, he must get it accepted, and he must get it past the censorship and opposed norms of the valued groups.

Let me call the reader's attention here to a term which is likely to occur in some of the later papers of this series—"feedback." By feedback, we mean information that comes back from the receiver to the sender and tells him how well he is doing. When I talk to you, and you nod your head in approval, I assume the message is getting through and probably being accepted. There is a great deal of feedback in personal communication, very little in mass media communication; that is one reason why it is easier to explain something, or convince a person, face to face.

So far, we have been discussing the simplest unit of communication—the two-person group, the sender and the receiver. Communication within society is made up of complicated networks and long chains of senders and receivers. Nothing is more characteristic of modern communication than these long chains—for example, the chains that bring news across the world. One very important characteristic of such chains is that every person on the chain except the first and the last is a gatekeeper. That is, he can pass on the message, or not pass it on, as he chooses. He can leave part of it out or add something to it. Therefore, he has great power over the message, and over the knowledge of everybody after him on the chain. When we consider how many gatekeepers intervene, for example, between a news event in Asia and its publication in a newspaper on the other side of the world, we are no longer surprised that errors or omissions sometimes occur.

What does communication do in society? It maintains the working relationships between individuals and between groups and nations. It engineers change, and it keeps strain at a tolerable level. Therefore, wherever there is pending change or trouble in society, there is a great deal of communication. When a group discovers it has a deviant member, it directs most of that communication to him until he returns to the fold or the cause is found to be hopeless. When a country decides it must industrialize, then it steps up its communication because the people must be informed and motivated.

Let us think of a primitive tribe, and how it must have used communication. It would post a watchman, to tell of danger or of opportunity—perhaps an enemy tribe approaching, or a herd of animals to be hunted for food. It would have a council to make decisions and see that they are carried out. It would have some way to store the wisdom of the tribe and some way to pass it on to the children and other new members; probably the elders would teach them the tribal customs, the mother would teach the girls to cook and sew, the father would teach the boys to hunt and fight. Then they would have entertainment—a bard to tell the old stories of the tribe, a minstrel to sing the favorite songs, dancers to perform the ritual dances.

These were, of course, not all the functions of communication in the tribe, but they were important ones. It is interesting to note that these still are the important functions of communication in society. We still need information on our environment, although now it is more likely to come from news services, newspapers, and radio. We still need communication machinery to get decisions made—to feed in the information, to find out the state of public opinion and try to achieve consensus, to make the decisions known and rally the people behind them. Now this is done by formal governments and through dialogues in groups and in the opinion media. We still need places to store the accumulated knowledge and wisdom of the race, and this is why we have libraries. We still need a way to socialize the new members of society, though now we do it chiefly through schools. And we still need communicated entertainment, although much of this, too, has been turned over to the mass media.

Beyond all these formal needs for communication, however, we have still another need. This is simply to maintain the everyday relationships of human beings who must live in proximity and must adjust to each other's needs and quirks and maintain a reasonably efficient and rewarding existence. What a wide variety of communication this includes—making love, borrowing a cup of sugar, saying "Good morning," playing a game, keeping the neighbor's children from trampling on one's flowers. The less efficient our communication is, up to a point at least, the less effective and pleasant these relationships are likely to be.

And just as interpersonal relations depend on the efficiency of communication, so do vastly larger things: the success of an advertising campaign, the adoption of a new product or practice, the changing of attitudes toward minority groups, the election of a president, the relationships between nations. All these are, in part at least, subjects for communication research.

Communication research, then, is concerned with how to be effective in communication, how to be understood, how to be clear, how people use the mass media, how nations can understand each other, how society can use the mass media to its greatest good, and in general how the basic process of communication works.

SUGGESTIONS FOR FURTHER READING

1 BERLO, DAVID. *The Process of Communication: An Introduction to Theory and Practice*. New York: Holt, Rinehart and Winston, 1960.

2 CHERRY, COLIN. *On Human Communication*. New York and London: published jointly by The Technology Press of M.I.T. and John Wiley and Sons, and Chapman and Hall, 1957.

3 HOVLAND, CARL I. "The Effects of Mass Communication," in Gardner Lindzey, ed., *Handbook of Social Psychology*. Boston: Addison-Wesley, 1954.

4 HOVLAND, CARL I., JANIS, IRVING, and KELLEY, HAROLD H. *Communication and Persuasion*. New Haven: Yale University Press, 1953.

5 KATZ, ELIHU, and LAZARSFELD, PAUL F. *Personal Influence*. Glencoe, Ill.: The Free Press, 1956.

6 KLAPPER, JOSEPH. *The Effects of Mass Communication*. Glencoe, Ill.: The Free Press, 1960.

7 SCHRAMM, WILBUR. *Mass Communication*. Urbana, Ill.: University of Illinois Press, 1960 (2nd ed.).

8 ———. *Process and Effects of Mass Communication*. Urbana, Ill.: University of Illinois Press, 1954.

2 THE THEORY OF COGNITIVE DISSONANCE

Leon Festinger

Human beings show a strain toward consistency. When they feel an inconsistency between what they know and what they have done, they often engage in rather unexpected communication behavior in order to reduce the discomfort that Dr. Festinger calls "cognitive dissonance." In this paper, he explains cognitive dissonance and describes some of the things human beings will do to get rid of it.

Professor Festinger is one of the best-known social psychologists in the United States. He was born in 1919 in New York, took his undergraduate studies at the College of the City of New York, and his graduate work and Ph.D. at the University of Iowa, where he was a student of Kurt Lewin. He has taught at the Massachusetts Institute of Technology, the University of Michigan, the University of Minnesota, and, since 1955, at Stanford, where he is a professor of psychology. His best-known work is his book, The Theory of Cognitive Dissonance, *published in 1957. He has also published* When Prophecy Fails, *a study of a religious group that predicted the world would end on a certain day in 1953, a handbook of* Research Methods in the Behavioral Sciences, *and a number of articles in the psychology journals.*

The relationship between what a person knows and how he acts is not a simple one. By and large, of course, people act in ways which are consistent with what they know. If a person is aware of some danger, he is usually cautious; if he knows one restaurant is better than another, he will eat in the better one, and so on. Frequently, however, inconsistencies occur between how a person acts and what he knows. We will here concern ourselves with some of the psychological consequences of such inconsistencies.

Let us start by considering some everyday kinds of behavior. At one time or another everyone has observed someone trying to justify his behavior. A student who has studied very hard for an examination may tell others what an important examination it really is. A parent who decides to send his child to some private school may overwhelm others with his description of the wonderful advantages of that particular school. A young man who spends more than he can afford to take some girl out to dinner may, on the next day, talk with enthusiasm about what a wonderful girl she is.

How may we explain and understand such behavior? Someone might say that it is very easy to understand on the basis of common sense and of well-accepted ideas concerning motivation and behavior. If the student had *not* thought the examination was very important, he would not have studied so hard; if the parent had *not* thought the private school was wonderful, he would not have decided to send his child there; if the young man was *not* enthusiastic about the girl, he would not have spent so much money to take her to dinner. All this seems very plausible and is undoubtedly true. There is, however, an additional question to ask. Why should these people talk about it so much?

The possibility exists that the action was taken with insufficient justification and, after the action has been taken, the person is trying to find *additional* justification for it. Let us see, theoretically, how and why this might be. Examine the information, opinions, and beliefs of, for example, the young man on the date. While his enthusiasm about the girl fits well with his behavior, his knowledge concerning his financial situation does not fit at all with what he is doing. This latter relationship between his action and his knowledge concerning his finances is of particular interest to us. We will call such relationships dissonant ones. The dissonance between this information and his behavior may be bothersome enough so that he tries to lessen its impact by magnifying those aspects of the situation which do fit with his actions.

One might, on the basis of this reasoning, hazard the following general theoretical proposition: Any time a person has information or an opinion which considered by itself would lead him not to engage in some action, then this information or opinion is dissonant with having engaged in the action. When such dissonance exists, the person will try to reduce it either by changing his actions or by changing his beliefs and opinions. If he cannot change the action, opinion change will ensue.

This psychological process, which can be called dissonance reduction, does explain the frequently observed behavior of people justifying their actions. The question arises, however, as to whether or not the hypothesized process of dissonance reduction does occur and, if so, under what conditions. In other words, how can we experimentally test this theory concerning dissonance?

In principle, the requirements for an experimental test are rather easy to specify. If we could deliberately create a specified dissonance between what a person knows and some action he has taken, and if we could measure his information and opinions relevant to the action both before and after the action was taken, we would then be able to determine whether or not the hypothesized dissonance reduction occurred.

In practice, of course, such well-controlled experimental situations are not easy to devise and carry out. The major problem is how to create, experimentally, a given dissonance between something the person knows and some action he takes. In the course of our research on the theory of dissonance, a number of means of carrying out such experiments have been devised. Perhaps the best way to present a description of the validity of the theory, and of the varieties of human behavior which it helps to understand, is to describe briefly a number of these experiments.

It will be recalled that, theoretically, dissonance is said to exist between some information and a given action if, considering this information all by itself, it would ordinarily lead a person not to engage in the given action. If this is true, it leads one to the idea that such dissonance must exist after a person has made a decision between two alternatives which are both rather attractive. After he has chosen one of the alternatives, everything he knows concerning desirable aspects of the rejected alternative are dissonant with the action he has taken, that is, the decision he has made. If this is true, and if as a result dissonance-reduction processes follow, one would expect that after such a decision the person increases the justification for his action by persuading himself that the alternative he chose is even more attractive than he had originally thought.

An experiment to test this specific derivation from the theory was conducted by Jack Brehm in the following way. He asked the persons in his experiment to rate each of a number of objects for attractiveness and desirability. Then, as an indication of his appreciation to them for participating in the experiment, he gave each of them a choice between two of the objects; whichever they chose they could keep. After they had made their choice, he asked them once more to rate the attractiveness of each of the objects.

The two objects between which they were allowed to choose were carefully selected by the experimenter. For half the people in the experiment, these two objects were selected so as to be close together in attractiveness according to the person's initial rating of them. For the other half, the objects were selected so as to be rather far apart in initial attractiveness. When the two objects were close together in attractiveness, there would, theoretically, be considerable dissonance following the decision, since the person knew many desirable aspects of the rejected alternative. When the two objects were far apart in attractiveness, there would, of course, be little dissonance following the decision. The results showed that when the rejected alternative is attractive and considerable dissonance results from the decision, the chosen alternative is judged to be more attractive after the decision than it was before. When there is little dissonance following the decision because the rejected alternative is not very desirable, the chosen alternative does not become more attractive. Thus, the data tend to confirm the existence of the process of dissonance reduction.

Another study related to the one just described may be of interest as an indication of how dissonance reduction after a decision relates to communication processes. Ehrlich and others did a study of the extent to which people who had recently purchased new automobiles read advertisements about new automobiles. The basis for the study was the idea that usually, before making a purchase of a new car, people consider various different kinds of cars. Thus, the actual purchase represents a decision, and dissonance should exist because of all the attractive features of those cars which they considered but did not buy. Further, in the process of attempting to reduce dissonance, they should be eager to see or hear information which informs them that the car they have recently purchased is, indeed, a very good car. Since this is precisely what advertisements do say, it was expected that persons who had recently bought a new car would show a strong tendency to read material advertising the car they had just purchased.

The investigators found this to be true. People were interviewed within four weeks after buying a new car. The data showed that in the preceding week they had read many more advertisements concerning the car they had just purchased than other automobile ads. Persons who were not very recent buyers of new automobiles did not, when interviewed, show any such tendency in their reading of advertisements. In other words, after making a decision, people look for, and usually find, ways to magnify their justification of their behavior.

One would, of course, also expect these dissonance-reduction processes to occur in situations other than those represented by recently made decisions. One very interesting type of situation in which dissonance should exist, and where we would expect dissonance reduction, is that in which a person is induced to say something publicly which does not agree with his private opinion. This kind of situation is, of course, rather frequent. For one or another reason, perhaps to obtain some reward such as getting a job or merely having the other person like him, or perhaps to avoid some unpleasantness, a person may say something which, because his true private opinion is different, he would ordinarily not say.

Let us analyze this type of situation a bit further. After a person has made the public statement which disagrees with his private opinion, there are two major classes of information he has that are relevant to this action which should be considered. The information the person has concerning the reward he has obtained for his action or the unpleasantness he has avoided by it is, of course, consonant with his having engaged in the action. The information he has concerning his own private opinion is, however, dissonant with the action he has taken, since, considering only his true private opinion and nothing else, he would certainly not have engaged in the action. If this analysis of the situation is correct, then we would expect that after a person had been induced to say something contrary to what he privately believes, there would be a process of dissonance reduction which would show itself by his seeking additional means of justifying his action. There are two major ways in which he could provide such additional justification for himself. He could magnify the importance of the already existing justifications by telling himself that the reward he obtained was really very great or that the unpleasantness he avoided would have been extremely unpleasant. He can also provide additional justification by changing his private opinion on the matter. If he changes his private opinion so that it corresponds more closely to what he overtly said, then clearly the dissonance is reduced.

We may illustrate this phenomenon by briefly describing an experiment conducted by myself and Carlsmith to test these implications of the theory of dissonance. In this experiment the plan was to induce all of the persons in the experiment to say something contrary to their private belief in order to obtain a reward. The reward to be used was money, so that it would be difficult for them to magnify or exaggerate the importance of the reward. Hence we expected that dissonance reduction would occur mainly through change of private opinion. In addition, half of the subjects were given a very large reward while the others were given a small one. The idea was that if one obtains a large reward there is already enough justification for the action so that less dissonance reduction through change of opinion should occur.

The experiment was done in the following way. Each subject in the experiment came to the laboratory and was asked to work for an hour on some manual tasks that were deliberately chosen to be fatiguing and boring for him. He was led to believe that the whole purpose was to study how people worked at manual tasks. Our real purpose, which the person did not know, was simply to provide each subject in the experiment with the same experience about which he would have a negative feeling.

After each person had finished working on the boring tasks, the experimenter pretended that the experiment was finished. He then attempted to enlist the cooperation of the subject in the following way. The experimenter told him that a girl was waiting who was to be the next subject in the experiment and also explained that part of the reason for doing the study was to see whether a person's expectations made any difference in how he or she worked on the tasks. For this reason, the experimenter said, he would like to hire him to help in the experiment and tell the girl who was waiting that the tasks were very interesting and a lot of fun. When the subject agreed to do this, the experimenter paid him. Some subjects were paid very little for doing this, while others were given a rather large sum of money. The subject was then taken into the next room where a girl, actually a confederate of the experimenter, was waiting. The subject then proceeded to tell the girl that he had just finished the experiment and that it had been very interesting and very enjoyable. The experimenter then thanked him and said good-by. The subject was later interviewed by someone else to measure how interesting and enjoyable he privately thought the manual tasks really were.

In other words, using the excuse of hiring the person to help him, the experimenter induced each subject to tell someone that the tasks were very interesting and enjoyable when privately he really thought they were dull and boring. The dissonance between these two pieces of information could be reduced by the subjects changing their private opinion. The results show that this does, indeed, occur, but only for those who are induced to engage in the action with very little justification initially. Those who were paid a large sum of money to tell the girl that the tasks were interesting continued to believe privately that they were dull and boring. These persons do not differ in their opinions about the tasks from subjects who simply worked on the tasks and were immediately asked their opinions of them. However, those persons who were paid relatively little to tell the girl that the tasks were interesting show evidence of having changed their private opinion to reduce dissonance. On the subsequent interview they rate the tasks as actually having been relatively enjoyable—the private opinions of these persons are considerably and significantly different from the opinions of those who were paid more money. In short, if a person is induced to act in a manner contrary to his private belief, and if the promised rewards or threatened unpleasantness which induce him to engage in the action are relatively small, he will later tend to bring his private opinion closer to what he has overtly and publicly said.

There are many more types of situations which ordinarily produce dissonance, and there are many more experiments which have been done which could be described. However, it is probably most useful to conclude with some discussion of the relation between the existence of dissonance and reactions to communication and influence attempts. Consideration of the matter leads one quickly to the idea that if a given dissonance exists, and if a person is trying to reduce the dissonance by changing some opinion which he holds, then that person will be very receptive to communications attempting to influence him in that direction. Likewise, of course, we would expect him to be very resistant to any influence attempt which would push him in the opposite direction.

Let us take an example to make this point clear. There are certainly many people who believe that one should brush one's teeth three times a day, after every meal. However, many, or perhaps most, of the people who believe this do not actually brush their teeth so often. Thus, dissonance exists between this belief and their behavior. We would, then, expect that such people would be readily influenced by a communication which told them that it is actually harmful to brush one's teeth too often, or by a communication which told them that a certain brand of tooth paste is so good that if one uses it, it is only necessary to brush one's teeth once a day. Either of these communications, if accepted and believed, reduces the dissonance which exists. On the other hand, if one attempted to persuade such a person that one should really, for adequate care of teeth, brush them five times a day, we would expect the communication to be resisted—the person would not be influenced. Clearly, if the person accepted this communication, it would merely increase the dissonance between his belief and his behavior. Without going into any detail, it may be stated that William J. McGuire reports an experiment which shows this effect very clearly. Persuasive communications are very effective when they reduce dissonance and quite ineffective if being influenced would simply increase dissonance. (This experiment is reported in my book, *A Theory of Cognitive Dissonance*.)

The existence of dissonance not only has an effect on a

person's receptivity to influence attempts but also has an effect on the initiation of communication and influence processes. Opinions which a person holds are usually not very easily changeable and so, in attempting to reduce dissonance, a person may seek to obtain social support from others for some new opinion which he would like to accept. Let us illustrate this by returning to the example of the person who does not brush his teeth as frequently as he believes he should. Let us further imagine that this person is not fortunate enough to have someone tell him that brushing the teeth is really unnecessary. Let us further imagine that his attempts at persuading himself of this are ineffective—all the experts say otherwise. This person, in an attempt to find others who believe what he would like to believe, namely, that frequent brushing of the teeth is unnecessary, may initiate conversations about it with people he thinks are likely to hold the opinion that he himself would like to accept. If he can find enough such people, or if he can persuade others of this, then he can perhaps acquire enough social support so that he himself may now accept the opinion as correct and thus reduce the dissonance between his beliefs and his behavior.

This point, of course, brings us back to the examples with which we started this discussion, that is, instances which we have all observed of people who try to justify their actions by telling others about them in ways which seem overenthusiastic. If our theory of dissonance and dissonance reduction is correct, we would see such behavior as attempts to obtain social support for the additional justifications which would reduce dissonance. We have seen that this theory does have validity. Laboratory experimentation has shown that the effects the theory would lead us to expect do indeed occur. When dissonance exists, dissonance-reduction attempts do occur.

There are, of course, problems in interpreting behavior in uncontrolled situations. In an experiment one can interpret the psychological situation rather clearly because the experimenter has created it under relatively controlled conditions. One should, however, be cautious about casually assuming the existence of dissonance in ordinary uncontrolled situations, or about assuming that certain forms of behavior are always oriented toward dissonance reduction. After all, if a young man takes a girl out to dinner and the next day tells what a wonderful evening it was and what a wonderful girl she is, it may merely be that he is giving a perfectly realistic account of the situation.

SUGGESTIONS FOR FURTHER READING

1 BREHM, JACK W. "Post-decision Changes in the Desirability of Alternatives." *Journal of Abnormal and Social Psychology,* 1956, *52,* pp. 384–389.

2 BREHM, JACK W., and COHEN, ARTHUR R. *Explorations in Cognitive Dissonance.* New York: John Wiley & Sons, 1962.

3 FESTINGER, LEON. *A Theory of Cognitive Dissonance.* Evanston, Ill.: Row, Peterson, 1957.

4 KATZ, DANIEL, ed. *Public Opinion Quarterly,* special issue on "Attitude Change," Summer, 1960 (particularly pp. 280–365).

5 ROSENBERG, MILTON J., HOVLAND, CARL I., *et al. Attitude Organization and Change.* New Haven: Yale University Press, 1960 (particularly pp. 164–197).

3 AN EXPLORATION INTO SEMANTIC SPACE

Charles E. Osgood

We all know what a word or a picture means to us, but how can we tell someone else clearly and accurately what it means to us? And how can we compare what it means to us with what it means to someone else? In other words, is there a scientific way to measure meaning? Dr. Osgood has been studying this problem and has developed some scientific yardsticks for connotative meaning; he describes in this paper some of the things he has done and the conclusions he has reached.

Dr. Osgood was born in 1916 in Massachusetts, earned his B.A. at Dartmouth College in 1939, and his Ph.D. in psychology at Yale in 1945. From 1942 to 1945, he served as instructor in psychology at Yale. After military service he taught at the University of Connecticut from 1946 to 1949. In 1949 he was appointed to the faculty of the University of Illinois, where he became a professor of psychology, and in 1955 he was appointed to succeed Wilbur Schramm as Director of the Institute of Communications Research. He is best known for his book, The Measurement of Meaning (1957), in which he describes and analyzes exhaustive research with the semantic differential, a research tool which he developed for measuring meaning. He is also the author of a text, Method and Theory in Experimental Psychology, and of a number of articles and monographs. He is president of the American Psychological Association.

For the past ten years or so, a group of us at the University of Illinois have been working on the theory and measurement of meaning. The early phases of this research, which are reported most completely in our book, *The Measurement of Meaning,* offered evidence for a reasonably stable set of dimensions within which meaningful judgments are made by American college students. In recent years, our efforts have been devoted to studying the commonness of this semantic system. But before discussing this work, let me provide some context by briefly describing the measurement model.

I will begin by asking the reader to do the impossible—to imagine a space of some unknown number of dimensions. This will be our hypothetical "semantic space," and we may explore it by analogy with the more familiar color space. Like all self-respecting spaces, this one has an origin, which we define as complete "meaninglessness" (like the neutral gray center of the color space). If we think of a word as being some point in this space, then its meaning could be represented by a vector from the origin out to that point: the length of the vector would index the "meaningfulness" of the word (like saturation in the color space), and its direction would index the "semantic quality" of the word (like hue and brightness in the color space). Furthermore, distance between the end points of any two vectors in this semantic space should index the "meaningful similarity" of the words thus represented.

But to talk about "direction" in any space, we need some reference coordinates. One more analogy with the color space will prove useful to us: just as complementary colors are defined as points equidistant and in opposite directions from the origin in the color space, which when mixed together in equal proportions cancel each other out to neutral gray, so may we conceive of *verbal opposites* as defining straight lines through the origin of the semantic space and canceling each other out to meaninglessness when "mixed." Imagine now a whole set of different straight-line "cuts" through the space, each defined by a pair of opposites: we might have a person indicate his "meaning" of a concept by a series of binary decisions—it is *beautiful* (not ugly), *soft* (not hard), *quick* (not slow), and so on. If these "cuts" were at right angles, and hence independent of each other—an assumption that demands justification, of course—then each binary decision would reduce uncertainty about the location of the concept by half. Or, if each straight-line "cut" were scaled into seven discriminable steps, as we have done in our work, then each decision would reduce uncertainty of location by six-sevenths, and only three "cuts" would yield a space of 343 discrete regions.

But we still have the problem of reference coordinates. Is the up–down, north–south, and east–west of the semantic space to be completely arbitrary, or is there some "natural," built-in structuring of the space analogous to the gravitational and magnetic determinants of geophysical space? This question is an empirical one, and the logical tool is some variant of factor analysis. We need to take a large and representative sample of dimensions defined by verbal opposites, determine their intercorrelations when used in judging the meanings of a representative sample of concepts, and then see if they do fall into "natural" clusters or factors that can serve as reference coordinates. In the past decade we have made more than ten such factor analyses of meaningful judgments by American college students, and over and over again we keep finding the same three dominant factors or dimensions: an Evaluation factor (represented by scales like *good–bad, pleasant–unpleasant,* and *positive–negative*), a Potency factor (represented by scales like *strong–weak, heavy–light,* and *hard–soft*), and an Activity factor (represented by scales like *fast–slow, active–passive,* and *excitable–calm*).

The problem before us now is this: How common is this semantic framework across different people doing the judging and across different kinds of concepts being judged? Is it limited to Americans speaking the English language, or is it shared by all humans regardless of their language or their culture? Is it the same for all concepts, be they aesthetic or political, familiar or unfamiliar, words or pictures? Let me anticipate our conclusion from a large number of studies— we find that the evaluation–potency–activity system is remarkably stable across people but quite unstable across the concepts being judged. We will want to inquire into *why* this should be so. But first, some evidence.

When a group of people judge a set of concepts against a set of scales, representing what we call a "semantic differential," a cube of data is generated. Each cell in this cube represents how a particular person judged a particular concept against a particular scale, using a number from 1 to 7. For example, in one cell we might have a number 7, this being Sally Smith's judgment of the concept "tornado" against a *slow–fast* scale (indicating that she thinks of "tornado" as *extremely fast*). In the next cell down in the cube we might find a number 4, this being Sally Smith's judgment of "tornado" in terms of *honest–dishonest* (the number 4 showing that she feels neither one way nor the other on this scale). Each person, as a subject, is a slice of this cube from front to back; each concept being judged, like "tornado" or "my mother," is a slice of the cube from left to right; and each semantic scale is a horizontal slice or row from top to bottom. Now, in analyzing these data we usually are interested in the correlations between scales—that is, in determining how the semantic dimensions cluster together—but we can run these correlations either across subjects or across concepts, and we can do it either for all subjects or concepts as a group or for individual subjects or concepts. As can be seen, there are many ways we can slice our semantic cake, and each method of slicing serves to answer a different kind of question.

Let us look first at the question of *generality of factor structure across people*. Within the English-speaking American culture, we have made many comparisons between *groups* of people—between old people and young people, between males and females, between students exposed to a new kind of course in international relations and those given a traditional course, between Republicans and Democrats, and even between schizophrenic patients and normal people. I can summarize the results of all these comparisons very simply: in no case have we found significant differences in the basic semantic factors. Note carefully that this *does not* indicate that the meanings of particular concepts were the same—they were not (Republicans have a very different meaning for "Harry Truman" as a concept than do Democrats, for example!). What this *does* indicate is that the semantic framework within which meaningful judgments are made is constant—the scales correlate with each other in the same ways for these different groups, despite the variations they show in the locations of particular concepts within the frame.

But the most critical test for generality of these semantic factors clearly would be *between people differing widely in both language and culture*. We have already made a number of cross-cultural comparisons—involving Japanese, Koreans, Greeks, and Navajo, Zuni, and Hopi Indians in the American Southwest—and the similarities in factor structure have been striking. But, for the most part, these studies have involved simply translating English scales into the various languages, and we are open to the criticism that we have forced people of other countries to operate within the limits imposed by an American English factor system. However, we do have one study done completely independently, by the Marketing Center Company in Tokyo, in which the same general factors appeared, and we are now in the middle of a large-scale cross-cultural study, involving some twelve countries and testing conditions as ideal as we can devise. I would like to describe this research project—its methods, its results to date, and its purposes.

With the help of cooperating social scientists in each country—without whom this type of research could not be done—we are collecting data in Japan, Hong Kong, India (Hindi in New Delhi and Kannada in Mysore), Afghanistan, Iran, Lebanon, Yugoslavia, Poland, Finland, Holland, Belgium, and France, along with the United States as a comparison base. We start with a list of one hundred familiar concepts that have been selected by linguists and anthropologists as being "culture-fair," and that have survived a stringent back-translation test with bilinguals, for all of the six language-families represented. This is the only point where translation is involved and could influence the results. From this point on, everything is done in the native language and with native monolingual subjects in each country.

The first step is to have one hundred young high-school boys in each country give the first qualifiers (adjectives, in English) that occur to them when each of the concepts is given as a stimulus—for example, to the word "tree" one boy might say "tall," another "green," another "big," and so forth. This basketful of 10,000 qualifiers (100 subjects times 100 concepts) is shipped to the University of Illinois, where, using IBM and ILLIAC high-speed computers, we determine a rank order of these ways of qualifying, in terms of total frequency of usage, diversity of usage across the hundred concepts, and independence of usage with respect to each other. We already have these rank-frequency-diversity measures for nine countries; they are not only highly similar in statistical properties, but when the ranked qualifiers are translated into English and then correlated with both English and each other, the correlations are all significantly positive. In other words, the dominant ways of qualifying experience, of describing aspects of objects and events, tend to be very similar, regardless of what language one uses or what culture one happens to have grown up in.

The second step in each country is to take the fifty highest ranking qualifiers, elicit their common opposites so as to make scales like *good–bad* and *big–little* out of them, and then have a new group of one hundred young men judge each such scale against every other of the fifty scales—to what extent is *good* either *big* or *little,* to what extent is *big* either *happy* or *sad,* and so on. This new basketful of data is shipped back to Illinois, where we do the correlations and factor analyses that represent our first test of the structure of the semantic space. For the six countries that have been carried to this stage, I can report that the first two factors are definitely, as expected, Evaluation and Potency; the third factor is more variable across countries but seems to have at least the "flavor" of Activity (semantic properties like *hot, fast, young,* and *noisy* keep appearing). We hope that the third step will clarify this situation. Here we will have yet another group of similar subjects judge the original hundred culture-fair concepts against the fifty semantic scales, correlate each scale with every other as used in actually judging concepts, and do another factor analysis. We are only beginning this phase of the research, so I cannot give any results as yet.

What is the purpose of all this busywork in many lands and many tongues? The first, purely scientific, purpose is to demonstrate that human beings the world over, no matter what their language or culture, do share a common meaning system, do organize experience along similar symbolic dimensions. In contradiction to Benjamin Lee Whorf's notion of "psycholinguistic relativity"—according to which people who use different languages *must* perceive things differently, think differently, and even create different philosophies—here may be at least one aspect of human symbolic behavior that is universal. A second, more practical purpose of this research is to develop and apply instruments for measuring "subjective culture"—meanings, attitudes, values, and the like—instruments that can be shown to be comparable across differences in both language and culture. The demonstration of common semantic factors—if indeed they can be demonstrated—makes it quite feasible to construct efficient "semantic differentials" for measuring the meanings of critical concepts cross-culturally, with reasonable confidence that the yardstick is something better than a rubber band. Ultimately, it would be my hope that both the demonstration of a shared semantic framework and the application of semantic measuring instruments would contribute to better international communication and understanding.

Let us flip the coin over and ask about the generality of semantic factor structures across the concepts being judged. The reader will recall that the cube of data generated when a group of subjects judges a sample of concepts against a set of scales makes it feasible to compute separate correlation matrices for each concept "slice" and factorize such matrices. In what we refer to as our "Thesaurus Study"—because the adjectives were sampled from that source on a rational, representative basis—twenty different concepts, like "foreigner," "knife," "modern art," "debate," and "hospital," were judged against seventy-six scales by one hundred college subjects.

Now, imagine the twenty separate correlation matrices for the different concepts lined up as a deck; if we go through the deck at the point of intersection of a particular pair of scales (for example, of *sober–drunk* vs. *mature–youthful*), we will isolate twenty *r*'s, one for each concept. If scale relations were reasonably constant over concepts, then we would expect only minor variations within such rows of correlations—but this proved *not* to be the case. Correlations were found to vary as much as from +.60 to —.60 in the same row. A couple of examples will serve to suggest what is happening: *sober* goes with *youthful* for the concept "dawn," but *sober* goes with *mature* for the concept "United Nations"; *pleasurable* goes with *feminine* for the concept "mother," but *pleasurable* goes with *masculine* for the concept "Adlai Stevenson." It would appear that the nature of the concept being judged exercises a restriction on scale meanings. What about the correspondence of factors derived from such single-concept matrices? Here the picture is better: something identifiable as an Evaluation factor appeared for each concept, and it was usually the first in order of magnitude; something identifiable as a Potency (or Dynamism) factor appeared for all but two concepts; but other factors varied in most inconsistent ways.

This instability of scale relations and factors across concepts contrasts sharply with the stability we have found across people. This shows up most clearly in studies where both types of generality can be compared. In one experiment designed specifically to get at this problem, college girls in both Japan and the United States judged three different classes of concepts—patches of color, simple line drawings, and abstract words like *love* and *peace*—against a 35-scale form of translation-equivalent semantic differential. Separate scale-by-scale correlation matrices and factor analyses were run for each of the six combinations of two subject groups and three concept classes. Now, if our hypothesis—that semantic systems are more stable across people than across concepts—holds, then factorial similarities should be higher when Japanese and American girls judge the same materials (both judge colors, both judge forms, and so forth) than when the same group judges different materials. This was true in every case. Even the salience of the three major factors shifted in the same ways for both Japanese and Americans—Activity is the dominant factor in judging colors for both groups, Potency tends to be the dominant factor in judging line forms, and Evaluation is clearly the dominant factor in judging abstract words for both groups.

So much for evidence. In the lines remaining to me I would like to speculate a bit on the "why" of these observations. Why do we find such wide generality of the evaluation-potency-activity framework across people, both within and between languages and cultures? And why, given generality across people, do we find such lack of generality in semantic structure across the classes of concepts they judge? While I certainly don't know the answers to these questions, I do have some hunches that I'd like to set down briefly here.

First, I must confess that when we began this research over ten years ago, I had expected the major dimensions of the semantic space to reflect the ways in which our sensory apparatus divides up the world. This was in flat contradiction to my own behavioristic theory of meaning, in which the semantic components should be *response-like* in character. Well, the accumulating facts have proven my expectation wrong and my theory at least "righter"—the dominant factors of Evaluation, Potency, and Activity that keep reappearing certainly do seem to have a response-like character, seemingly reflecting the ways we can react to meaningful events rather than the ways we receive them.

But these major factors also seem to have an *affective* as well as a response-like character. The similarity of our major factors to Wundt's tridimensional theory of feeling—pleasantness, strain, and excitement—has been pointed out to me. And, as a matter of fact, we have done a number of experiments on the meanings of facial expressions, coming out with Pleasantness, Control, and Activation as three factors which seem pretty much to exhaust the semantic space of facial communication. The similarity between these factors in emotional communication and those found in our more general linguistic studies suggests that the latter may also have their grounding in the affective reaction system.

Let me speculate a bit further and suggest that the highly generalized nature of the affective reaction system—the fact that it is independent of any particular sensory modality and yet participates with all of them—is at once the reason why Evaluation, Potency, and Activity appear as dominant factors *and* the psychological basis for metaphor and synesthesia. It is *because* such diverse sensory experiences as a *white* circle (rather than black), a *straight* line (rather than crooked), a *rising* melody (rather than a falling one), a *sweet* taste (rather than a sour one), a *caressing* touch (rather than an irritating scratch)—it is because all these diverse experiences can share a common affective meaning that one easily and lawfully translates from one sensory modality into another in synesthesia and metaphor. This is also the basis for the high interscale correlations which mathematically determine the nature and orientation of general factors. In other words, the "common market in meaning" seems to be based firmly in the biological systems of emotional and purposive behavior that all humans share.

And now, finally, what about the lack of generality of semantic factor structure across concept classes? All of the evidence we have clearly indicates that there is *interaction be-*

tween concepts and scales in the process of semantic judgment. What are the implications of this? For one thing, this means that from the standpoint of applied semantic measurement these can be no such thing as "*the* semantic differential." So, for particular concept classes we will need to construct appropriate differentials, and in the area of personality measurement we have already made a start. From the standpoint of psycholinguistic theory, the fact of concept/scale interaction invites fresh speculation about how it operates—and therefore a host of new experiments, most of which are far from fully conceived.

Let me refer, rather arbitrarily, to the aspect of "meaning" tapped by the major factors of the semantic differential as the "connotative meaning" of signs—this is the universal aspect based on the biology of the affective system. There is another aspect of "meaning" which deals with the elaborate set of essentially arbitrary correlations between linguistic and non-linguistic events—for example, between the noise "apple" in English and the visual perception of an *apple* object—and this I shall refer to as "denotative meaning." This is the meaning of "meaning" with which linguists and philosophers have been primarily concerned. I assume that these arbitrary correlations are mediated by the sensory and motor discrimination systems of the brain, regions where lesions may produce various types of aphasia. Both of these biological systems—the affective, energizing system and the sensory-motor discrimination system—are integrated in ordinary behavior, and I believe that integration of the same systems in language behavior is one reason for concept/scale interaction.

If I were to ask the question, "Is a baby *large* or *small?*" you would undoubtedly say, "small." After all, within the class of human organisms, a baby *is* "a small one." Within the class of nails a spike is "a large one." I think the semantic differential technique, in which a single stimulus is judged successively against a series of different scales, is one which tends to draw out these intraclass connotations of signs. In all other psychophysical methods with which I am familiar, even the so-called absolute judgment method, many different stimuli are judged successively against a single scale—for example, in judging weights or in scaling the loudness of tones. Note that if I ask you to compare "baby" and "spike" in terms of size, you immediately say that "baby" is "larger," now disregarding the intraclass connotations of these objects.

What has all this to do with concept/scale interaction? I think that the semantic differential is subject to what might be called "denotative contamination." The terms that define our scales have variable denotative meanings as well as their generalized affective connotation. The denotation of *masculine–feminine* is brought out by the concept "Adlai Stevenson," whereas its potency connotation is elicited by the concept "dynamo"; a concept like "lava" taps the denotation of *hot–cold,* whereas concepts like "jazz" and "festival" call forth the general connotation of *hot.* It is clear that if certain scales are denotatively relevant to certain classes of concepts, they will fall away from their usual affective factors and hence change the total structure. We have begun a series of experiments comparing the two basic judgmental methods —one concept at a time against many scales vs. many concepts against a single scale at a time—and I am hopeful that these experiments will help us disentangle denotative and connotative aspects of meaning.

Another probable source of concept/scale interaction I should mention in closing is what we call "cognitive interaction." This is the tendency for two simultaneously evoked meanings to change each other in the direction of compromise, presumably because the affective system can only assume one "posture" at a time. In making semantic-differential judgments, you first look at, and then "keep in mind," the concept being judged, while you go down the page placing appropriate check-marks on each scale. If the concept "mother," for example, is being judged and it has an intensely positive evaluation, this meaning should interact with those of the scale terms and cause them to become momentarily *more* evaluative. In mathematical terms, this means a general rotation of scales toward the dominant evaluative factor. In other words, each concept or concept class will tend to produce rotation of scales toward its own characteristic attribute in the semantic space. We now have some experiments on the planning boards in which we will try to predict the rotations of scales in the factor space from knowing the measured meanings of the concepts and the scale terms that are interacting.

This brings us to the end of a brief flight through semantic space. I shouldn't say the "end" because, like any other scientific exploration, this one has no clearly identifiable beginning and certainly no point of conclusion. We are always "in progress," with each experiment merely beckoning for more. Throughout this paper I have used the term "we" rather than "I"; this is because the research I have described has been the work of many people, not only at the University of Illinois but also in many other universities in many countries. I have not mentioned any of them by name because it would have been impossible to mention them all. But I do want to express my appreciation for their interest and their help.

SUGGESTIONS FOR FURTHER READING

1 KATZ, DANIEL, ed. *Public Opinion Quarterly,* special issue on "Attitude Change," Summer, 1960 (for relation of Osgood's congruence hypothesis to other theories of cognitive consistency).

2 OSGOOD, CHARLES E., SUCI, GEORGE J., and TANNENBAUM, PERCY H. *The Measurement of Meaning.* Urbana, Ill.: University of Illinois Press, 1957.

4 THE NEW "SCIENTIFIC" RHETORIC

Nathan Maccoby

Is there a set of principles by which we can predict whether a given communication will have a desired effect on a specified receiver? These principles are what the rhetoricians of ancient times tried to develop, and for centuries students learned their "rules of rhetoric." Now, communication research is making it possible to develop rhetorical principles, based on scientific investigation, about which we can be more confident than about some of the old principles. Chief figures in this development have been the Yale group around Carl Hovland. Dr. Maccoby discusses the Hovland group and some of their findings in the following paper.

Dr. Maccoby was born in London in 1912, was brought to the United States at an early age, and is a naturalized American citizen. He graduated from Reed College in Portland, Oregon, and earned a master's degree from the University of Washington and a Ph.D. from the University of Michigan. He was a member of the famous Army research unit headed by Carl Hovland and Samuel Stouffer during World War II. Later he was a research director for the Survey Research Center at the University of Michigan. He taught at Oregon State and at Boston University, where he became head of the Department of Psychology; since 1958 he has taught at Stanford, where he is Professor of Mass Communication and a member of the Institute for Communication Research. His specialty is the psychology of communication, in which field he has a long list of research articles and monographs.

Traditionally, rhetoric has been concerned with the art of persuasion. The ancient Greek sophists were so severely accused of dishonest intellectual acrobatics that the very word *sophistry* came to mean, as it still does, "trickery with words." Even Grote, the great defender of the sophists, characterizes Plato's attack on them as "sophistry." John Stuart Mill, the famous nineteenth-century British philosopher, said that the sophists have remained an object of scorn through the centuries because of the charge that they corrupted the youth of ancient Greece, which, Mill declared, is true also of Socrates, Plato, Aristotle, and all great teachers since. "Whatever encourages young men to think for themselves," said Mill "does lead them to criticize the laws of their country—does shake their faith in the infallibility of their fathers." Naturally, their fathers object.

Perhaps because of his very great reputation in other fields of knowledge, Aristotle did not suffer the fate of the sophists and is generally regarded as the founder of rhetoric as an academic discipline. His treatise bears the title *The Art of Rhetoric*. Plato, Aristotle's teacher, defined rhetoric as "the winning of men's minds by words." For Aristotle, rhetoric was "the faculty of discerning the possible means of persuasion in each particular case." For both, the approach to rhetoric was through psychology—to them, the science of the mind. Aristotle—and here he was unlike his teacher, Plato, as well as his other predecessors in ancient Greece—divorced ethics from science. He was concerned with the art of persuasion, the way the mind worked. He wanted an objective analysis of this process, unclouded by considerations of the good and the bad. He had to rely on the grossly primitive psychology of his day.

The new rhetoric is also concerned with the persuasive process. Like Aristotle's rhetoric, it is concerned with objective description and analysis of the processes of persuasion and, like Aristotle's, it is based on psychology. But unlike Aristotle's, it has a substantial body of information on human behavior from the modern science of psychology on which to draw. In fact, it might better be said that a substantial portion of the present-day psychological knowledge of human social behavior *consists* of the knowledge accumulated about the effects of persuasive communications.

While there have been isolated instances of psychological experiments in attitude change being conducted earlier, most of this knowledge has been acquired during the last quarter-century. A very great deal of it was conceived and carried out by the late Professor Carl I. Hovland, of Yale University, and his colleagues and students. Hovland's monumental work is for the most part reported in six volumes. Because these constitute reports of such an important portion of the work in the new rhetoric, I shall list them: *Experiments in Mass Communications*, Hovland, Lumsdaine, and Sheffield; *Communication and Persuasion*, Hovland, Janis, and Kelley; *Order of Presentation in Persuasion*, Hovland *et al.*; *Personality and Persuasibility*, Janis, Hovland, *et al.*; *Attitude Organization and Change*, Rosenberg, Hovland, *et al.*; and *Social Judgment*, Sherif and Hovland.

The first volume reports a series of studies carried out by Hovland and his associates in the United States Armed Forces during World War II. A series of films about the origins and objectives of the Allies' great struggle with Germany and Japan were prepared and widely shown. These films portrayed not only the role of the United States in World War II but also those of its allies, Britain and the Soviet Union. In a series of carefully designed experiments, including appropriate control groups who did not see the films, Hovland and his associates measured the effects of these films on troop information and attitudes. These first studies were straightforward evaluation experiments designed to measure relative effects of particular communications. Later studies undertook to measure relative effects of controlled variations. For instance, relative effectiveness of two different ways of presenting a point of view were experimentally compared. In one instance, the opposition's position was first presented and then refuted. In the other instance, only the point of view being advocated and the arguments for it were presented. In brief, the experimenters found that both kinds of presentation were about equally effective but that their effectiveness varied for different people. Those who were initially inclined to support the position being advocated were more likely to be influenced by the one-sided communication, while the two-sided communication was more effective for those who were opposed at the outset.

This experiment led to a whole series of subsequent ones. For instance, Lumsdaine and Janis found that, although they were able to confirm the above findings in every respect, a rather interesting thing happens when the two groups are later exposed to a countercommunication that argues against the position taken in the original communication. Again there is no difference in the proportion of people who change in the direction originally advocated, whether they were given a one-sided communication only, or a two-sided one. However, those given a two-sided communication proved to be highly resistant to the effects of a subsequent communication advocating the opposite point of view, while those given only the one-sided message were not.

Apparently, presenting the other side's position *first* somehow immunizes people, at least to some degree, against the effects of subsequent counterarguments. McGuire has recently discovered that this immunization works even when the particular counterarguments used in the later communication are different from those used in the initial training. Crane, more recently still, has found that even the mere mentioning that there are counterarguments, without the actual stating of any, has some immunizing power against the effects of subsequent countercommunications. These findings have been obtained on varying populations with a whole host of topics ranging from dental hygiene to world affairs. They are based on studies meeting the rigorous canons of scientific experimentation.

The second volume of the series deals with a great many experiments organized around three topics: first, the communicator, that is, *who* is doing the communicating; second, difference in the content of messages; and finally, variations in the characteristics of receivers of communications. In addition, there is continuing theoretical speculation about the *nature* of the *attitude-change* process. However, most of that comes later.

The first research on the effects of systematically varying the credibility of the source, or the communicator, is reported by Hovland and his student, Walter Weiss. An interesting finding occurs. As expected, when the communicator is a highly respected source, the communication is more persuasive than with a less prestigeful communicator. However, this difference dissipates over time. When subjects initially exposed to the same communication but from attributed sources varying in credibility are tested some four weeks later, those with the high source regress toward their original positions. However, those whose source was less credible show a delayed change upward or a "sleeper" effect. The result is that the difference due to source dissipates over time. Hovland and Kelman later found that this phenomenon is due to the fact that although the content of the message is retained over time, the source of the communication is not. Reinstatement of the source without any new presentation of the contents of the communication restored the original difference.

One of the noteworthy studies reported in this volume dealing with the systematic variation of message content is one done by Irving Janis and Seymour Feshbach. In this experiment the fear-arousing properties of the communication content were systematically varied, and the effects on attitude change and behavior compliance assessed. An attempt was made to induce secondary school children to adopt an approved method for brushing their teeth. Pupils were randomly assigned to one of four groups. Each of three groups was given a fifteen-minute illustrated lecture on the need for proper brushing of the teeth. One group, a control, received no lecture at all.

The three lectures differed from each other markedly in the emotional appeal of the arguments given for proper brushing. One was a very strong fear-arousal appeal: pain from toothaches and such afflictions as cancer, paralysis, and blindness resulting from extreme cases of bad teeth were discussed and illustrated. Having teeth pulled, cavities drilled, and other painful dental work was stressed. In the moderate fear-arousal lecture, none of these was stressed. Instead, mouth infections, sore swollen gums, decayed teeth, and cavities were discussed. These were also included in the first lecture. Of all these, only decayed teeth and cavities were stressed in the third, or minimal appeal, lecture.

All three lectures were effective but, surprisingly, the strong appeal was *less* effective than the more moderate presentations in inducing concern about decaying teeth or gums. This was also the case for the comparative effects of the talk on reported changes in tooth-brushing behavior. The minimal fear-arousing communication worked best; the strong one was so ineffective in this respect that no clear-cut effect on tooth-brushing behavior as compared with the control group was found. Furthermore, when the effects of a subsequent countercommunication were tested, the minimal fear-arousing condition was still the most powerful. An experiment recently reported by Robert Terwilliger and Janis confirms the explanation of these findings that strong fear raises strong defenses thus preventing attitude change.

Some of the first work on the role the receiver of communications plays in attitude change is also reported in this volume, although most of this work is reserved for a later report. The question that Janis and his colleagues raise here is: Are there some people who are generally persuasible, who tend to be swayed by any persuasive communication regardless of the content of the message? They point out that although a good deal of research had been previously devoted to the study of personality factors predisposing some individuals to be highly receptive to a given particular communication, the problem of *general* susceptibility had not yet been studied—persuasibility that instead of being topic-bound is topic-free. The authors also feel that research on persuasibility can throw light on the internal nature of the opinion-change process. Inferences about such mediating processes cannot be tested directly, but if external stimuli are presented which induce a change in certain kinds of people—mentally alert and well-functioning people, for example—and fail to induce change in mentally ill persons, some insights into the change process itself are at least suggested.

Much more work on personality factors relating to change-ability of opinions is reported in *Personality and Persuasi-bility*. This monograph reports two major sets of experiments. One of these is a much more extensive account of the personality correlates of general persuasibility; the other deals with the developmental aspects of persuasibility. Janis and Field did a basic study in this problem area by administering to the same subjects five different and unrelated persuasive communications. They found that, for the most part, the same people tended to be the ones who changed their minds as a result of the communication regardless of the issue involved. Godwin Chu is replicating this study in Tapei.

Although the results are by no means unequivocal, some personality variables do emerge as being related to general persuasibility. For example, among male subjects, low self-esteem is related to general persuasibility, but this is not the case for women. Similar results obtain for richness of fantasy. Rich fantasy is correlated positively with changeability for men but not for women. In fact, no statistically significant relationships occurred among women on any of the measures employed. Generally, women as a group are more persuasible than men. It may be that in our society socially prescribed roles differ sufficiently for the two sexes to produce differences in their basic ways of reacting to persuasive communications.

Two years earlier, in 1957, Hovland and his associates published a volume devoted to attitude-change experiments, all of which are organized around a single variable. Prior groups of studies had for the most part been omnibus collections of experiments. True, there had been a few series of experiments organized around single rubrics. The studies in prestige of the source, and the one-sided versus two-sided presentations of communications experiments are examples. But the group of studies reported in *The Order of Presentation in Persuasion* represented a conscious attempt to make organized units of problem areas. Actually, as we shall see, such a grouping did not result in any truly meaningful organization because the problem of order of presentation of arguments is not a sufficiently theoretically meaningful one. It remained for the next to the last volume in the series to report the tackling of more highly systematic organizations of research. Even with this limitation, however, some interesting and important problems in the effects of organizing communications on changes in attitudes are presented.

Some early work by F. L. Lund (1925) suggested that in ordering contradictory or opposing communications on a controversial issue, other things being equal, the side presented first was most likely to win the argument in the hearers' minds. Hovland and Mandell questioned the generality of this finding of Lund's and began by replicating Lund's earlier experiments but controlling for a methodological shortcoming of the earlier work. Only one of three tries successfully replicated Lund's finding of the greater potency of the first-presented order. In the other two replications, no consistent superiority of one order over the other occurred. Then, in a second series, a straight replication of Lund was carried out, but this time using issues that were more appropriate to the period (1949). This time none of the four issues tried gave primacy the superiority in persuasion. In fact, one gave recency a statistically significant edge. The authors then proceed to analyze the causes of the differences.

Lund had been the teacher of his subjects; therefore, whatever he presented first was likely to be accepted as true. But when he changed to the opposite position, his students must have felt that he was doing tricks with them and no longer believed him. Possibly, too, there was external compliance without genuine acceptance of what the teacher, who was going to read and grade their papers, had to say. Was there a difference between such compliance and true acceptance or commitment? This speculation led Hovland and his colleagues to do a further experiment on the role of commitment in the effects of order of presentation. Students were asked to write short paragraphs to be published, under their names, in a public affairs pamphlet to be circulated among their classmates and friends. When this "commitment" procedure preceded a communication, the change in attitude brought about by the message was less than that of a standard communication group. However, when the commitment procedure followed the communication, no such dampening effect of commitment took place. The authors conclude that when a person knows in advance that he is going to have to present and defend a position, he is less susceptible to a subsequent countercommunication.

Time does not permit detailed reporting of the many further experiments, but some of the other main findings are:

When two sides of an issue are presented successively by different communicators, the side presented first has no consistent advantage.

When contradictory information is presented by a single communicator, material presented first tends to be more effective than are communications presented subsequently. However, interpolation of other time-filling and unrelated activity between the two presentations can eliminate this primacy effect.

Presentation of information that has to do with need satisfaction after need arousal occurs is superior to the reverse order.

The influence of order is weaker with those people who have a strong desire for understanding (need for cognition).

Placing first the communication whose contents are highly desirable to the people involved is more effective persuasion than is the reverse order.

When an authoritative communicator plans to mention non-salient arguments contrary to the position he is advocating, it is more effective persuasion if he gives his own arguments first rather than second.

All of the above findings are documented by carefully controlled experiments. Finally, Anderson and Hovland present a mathematical model for representing the effects of order of presentation.

In April, 1960, just a year before Professor Hovland's untimely death at the age of forty-eight, the volume entitled *Attitude Organization and Change* was completed. Subtitled *An Analysis of Consistency Among Attitude Components*, it represents a major attempt to come to grips with several separate but not unrelated major theoretical approaches to attitude change, and to report a series of systematically controlled experiments designed to test hypotheses deriving from these theoretical efforts. Most, but not all, of these theoretical formulations are of the homeostatic or balance variety and are concerned with the effects of disruptions of cognitive consistencies. Most prominent of these is the theory of cognitive dissonance, which has been discussed more fully earlier in this book by its author, Professor Festinger. Briefly, it states that two cognitive elements are in a dissonant relation if, considering these two alone, the reverse of one element would follow from the other. Dissonance can arise not only from perceived logical inconsistencies but also from conflicting motives. When it does happen, there is postulated a drive to reduce it. Several ways of reducing it are possible, and these may be experimentally manipulated in any given situation. The theory has been applied more widely than merely to the effects of communications, but it has been specifically applied to communications effects by several experimenters, some of whose studies are reported in *Attitude Organization and Change*.

One of the most interesting formulations is that of Milton J. Rosenberg. This psychologist argues that when the cognitive and affective, or emotional, components of an attitude are in conflict, something has to give. He reasons that people seek congruence between their beliefs and their feelings toward an object, and from this he develops what he calls a "structural theory" of attitude dynamics. Rosenberg points out that while a number of studies have already demonstrated cases in which the affective component changed following cognitive change, few experiments have been done in which the cognitive component was the one changed to achieve congruence following a change in feelings. In a series of dramatic experiments, Rosenberg hypnotized subjects and suggested a radical change in feeling without giving reasons, plus an instruction to forget afterward that such a suggestion had been made. The results were quite startling. Not only did the changes in belief occur, but all sorts of arguments were invented to support these new beliefs. For instance, in one study hypnotized subjects who believed strongly in United States aid to foreign countries were told that, upon awakening, they would strongly oppose this principle of United States aid. They were completely convinced that they took their new positions as the result of hard reasoning. Rosenberg thus also stresses a homeostatic model with the factors in balance, the factors in this case being cognition and affect, with affect stressed as the modifier of cognition.

While Rosenberg is concerned with the role of drive produced by inconsistency in the cognitive and affective elements in promoting attitude change, William J. McGuire has taken a somewhat different approach. His approach is in terms of logical thinking and wishful thinking. A person's belief system is seen as being comprised of a series of interrelated propositions. In McGuire's model, when a series of propositions implies an illogical relationship or a logical inconsistency, wishful thinking is involved. Motivation to change one or more of the propositions so as to restore logic takes place when the wishful thinking is exposed. The Socratic method is employed as one of the ways of exposing such logical inconsistencies.

Jack W. Brehm, using the liking for a given vegetable and the estimate of its vitamin content, found that children committed to more eating through manipulation of rewards showed less lowering of their vitamin estimates from nonsupporting communications than did those committed to less eating, or than did control subjects. These results confirmed the prediction that commitment to a great deal of discrepant (dissonant) behavior will increase resistance to acceptance of a non-supporting communication and will increase acceptance of supporting information. Brehm concludes that when there is a feeling of choice in behavior at variance with other attitude components, dissonance will be created and attitude change will follow as a means of reducing this dissonance. Brehm and Arthur R. Cohen have recently published a volume reporting later studies entitled *Explorations in Cognitive Dissonance.*

The most recent volume in the series is *Social Judgment.* Some of the reported experiments had been done in the 1950's and had been previously referred to elsewhere by Hovland. Mainly, the work is concerned less centrally with the effects of communications on attitudes than it is in mapping out anchoring perceptions—that is, how extreme do people perceive a given attitudinal statement to be, and under what conditions do these perceptions vary? As a consequence of studying these problems, however, some attitude-change hypotheses emerge. Perhaps the main one has to do with what the authors call assimilation and contrast. When positions are seen as being close to one another, they tend to be perceived as more similar to one another than they actually are (assimilation), and when they are perceived to be distant from one another, they tend to be judged as even farther away from each other than they really are (contrast). The resulting hypothesis for attitude change is that when communications are perceived as being not very different from one's own actual position on an issue, they will be assimilated. When, however, the communication is perceived to advocate a quite different position, it will be rejected as being even more divergent than it really is from the recipient's own attitude (contrasted). Hovland, Harvey, and Sherif report experiments confirming this hypothesis. It should be noted, however, that the data on this problem are by no means all in as yet. Other experimenters, at least tentatively, seem to be getting differing results.

Each successive volume in the series contains an increasing amount of space devoted to suggestions for further research. In addition, although by far the major portion of the work in this field thus far has been done by Professor Hovland, his colleagues, and his former students, there are a number of other workers not mentioned in this discussion. The new scientific rhetoricians, far from confining themselves to the repeated discussions of the sophists, Socrates, Plato, Aristotle, and Quintilian, are finding, through the inventive formulations of theory and through submission to the impersonal fire of experimental tests of the hypotheses growing out of these formulations, an ever-widening scope of inquiry.

SUGGESTIONS FOR FURTHER READING

1 HOVLAND, CARL I., LUMSDAINE, ARTHUR A., and SHEFFIELD, FRED D. *Experiments on Mass Communication.* Princteon, N.J.: Princeton University Press, 1949.

2 HOVLAND, CARL I., JANIS, IRVING L., and KELLEY, HAROLD H. *Communication and Persuasion.* New Haven: Yale University Press, 1953.

3 HOVLAND, CARL I., *et al. The Order of Presentation in Persuasion.* New Haven: Yale University Press, 1957.

4 JANIS, IRVING, HOVLAND, CARL I., *et al. Personality and Persuasibility.* New Haven: Yale University Press, 1959.

5 KATZ, DANIEL, ed. *Public Opinion Quarterly,* special issue on "Attitude Change," Summer, 1960.

6 ROSENBERG, MILTON J., HOVLAND, CARL I., *et al. Attitude Organization and Change.* New Haven: Yale University Press, 1960.

7 SHERIF, MUZAFER, and HOVLAND, CARL I. *Social Judgment.* New Haven: Yale University Press, 1961.

5 PERSONALITY AS A FACTOR IN SUSCEPTIBILITY TO PERSUASION

Irving L. Janis

Some people, at some times, are more easily persuaded than others to do some things or to hold some belief. This paper by Dr. Janis summarizes what has been done to find out whether some people are in general more easily persuaded than others. Is it true, he asks, that some people are, by personality, more persuasible than others, and if so, what kinds of people are these?

A number of answers to these questions have come during the last ten years from the brilliant and productive group of psychologists who gathered around Carl Hovland, at Yale, in the years after World War II. Irving Janis was one of these. He was born in Buffalo, New York, in 1918, earned his first college degree at the University of Chicago, and in 1948 received a Ph.D. in psychology from Columbia. He was in governmental research during the war, and soon thereafter joined the Department of Psychology at Yale. He has now succeeded Hovland as head of the communication and attitude-change project at Yale, where he is a professor of psychology. His best-known books are Communication and Persuasion, *which he wrote jointly with Hovland and Kelley;* Air War and Emotional Stress; Psychological Stress; *and* Personality and Persuasibility, *the subject of which is the topic of this paper.*

It is a well-known fact that the same social pressures are experienced in different ways by different people. In other words, reactions to persuasion are determined not only by *who* says it and by *what* is said but also by the social and personality characteristics of the persons to *whom* it is said.

After a mass audience has been exposed to persuasive communications—such as controversial magazine articles, newspaper commentaries, political speeches on the radio, documentary movies, and political programs on television—which types of persons are most likely to change their personal beliefs and attitudes? Which types of persons tend to be most resistant? When we speak about "personality" as a factor in susceptibility to persuasion, it is essentially this problem with which we are concerned. Systematic research on personality characteristics and other predispositional factors should ultimately enable us to improve our predictions concerning the way that a given type of audience, or a given type of political leader or follower, will respond to new information and to powerful emotional appeals.

In this paper I shall describe some of the research studies by American social psychologists during the last decade on these problems. Many relevant studies have been conducted by my colleagues and myself at Yale University as part of a communications research project initiated by the late Professor Carl I. Hovland. Our experimental studies, which have been carried out in high school classrooms, in group meetings, and in other natural social settings, have usually been concentrated on various stimulus factors that are likely to determine the effects of a persuasive communication.

In general, our results bear out the work of many other social psychologists and sociologists in indicating that the *net effect* of mass media communications tends to be very limited, often consisting only of *reinforcing* pre-existing beliefs and attitudes. Attempts at producing major changes in social prejudices and political stereotypes generally meet with an extraordinarily high degree of psychological resistance, the sources of which we are just beginning to understand.

Thus, we have learned something about the general conditions under which resistance to change will be relatively high and low. We can describe some of the communication factors that help to produce a new outlook in an audience on those rare occasions when their resistances have been lowered sufficiently to permit at least a slight modification in their pre-existing attitudes. Here I am referring to such factors as the perceived role of the communicator, the types of emotional appeals used, and the order of presentation of positive and negative arguments.

In the course of studying such stimulus variables, we also began to examine *individual differences* within audiences exposed to the same persuasive messages, and gradually we began to accumulate some clues concerning the types of persons who were likely to be least resistant.

We have found it useful to take account of three different classes of personality characteristics which influence a person's responsiveness to persuasion: (1) his readiness to accept a favorable or unfavorable position on the particular *topic* that is being discussed; (2) his susceptibility to particular types of arguments and persuasive appeals; and (3) his overall level of susceptibility to any form of persuasion or social influence. I shall discuss the first two of these categories very briefly and then concentrate mainly on the last category, which involves a concept of "general persuasibility" that has only recently begun to emerge from factor analysis studies and related types of systematic research.

The first category has long been recognized as a major determinant of individual differences in responses to nationalistic propaganda and hate campaigns, as well as to educational programs designed to promote racial tolerance and international understanding. For example, there are many studies of authoritarian personalities which indicate that pro-democratic communications, intended to break down social prejudices, will be most strongly resisted by one particular type of personality, whose outstanding characteristic is a strong, latent need to displace hostility toward remote social targets—as manifested by symptoms of intense ambivalence toward parents and other authority figures, combined with inhibitions of normal sexual and aggressive activities. These personalities show an unusually intense interest in the so-called immoral behavior of people in foreign countries, or in other out-groups, and high anxiety about deviating from conventional moral standards.

Turning now to the second type of predispositional factor, we take account of the fact that if an audience is exposed to a series of persuasive communications on a wide variety of different topics and if many different types of emotional and rational appeals are used, there will be some persons who are influenced whenever a *certain type* of appeal is used, whereas others will not be responsive unless quite a *different* type of appeal is used. For example, some personalities are likely to respond favorably, and others unfavorably, to communication appeals that involve the arousing of guilt, shame, fear, or other strong emotions.

In one of our studies, we obtained evidence concerning individual differences in reactions to a strong, fear-arousing message that attempted to persuade adolescents to change their dental hygiene habits by playing up the threat of suffering from pain and disease. The chronically anxious persons— the ones who were most excitable and timid, who continually worried about dangers in daily life, who suffered from acute neurotic anxiety symptoms—did *not* turn out to be the ones who were most responsive, as might have been expected on a common sense basis. Rather, it seems that these anxious people were more likely than others to *reject* the communicator's conclusions and to show other unfavorable reactions, even though they were the ones most frightened by the threat appeal.

This finding highlights the need for systematic investigation of the apparently "common sense" rules of thumb that so often enter into the calculations of public health experts and government officials who attempt to modify the public's beliefs and attitudes on matters related to the threat of illness, disaster, war, or other recurrent sources of danger.

Differences in intellectual ability, as indicated by I.Q. and amount of schooling, must also be taken into account, especially when we try to predict reactions to various types of rational arguments in communications dealing with controversial issues that are in the focus of public attention. For example, one study indicates that persons with *low* intellectual ability and with relatively little education tend to be more influenced if the communications contain a *one-sided* presentation, limited solely to the arguments favoring the communicator's conclusion; whereas, better educated persons with high intellectual ability tend to be more influenced if the communication contains a *two-sided* presentation that includes at least a small amount of discussion of the other side of the issue.

Responsiveness to different types of emotional and rational appeals are also likely to be related to ethnic and national differences. Arguments and appeals that exert enormous influence in one country may fail to arouse any interest at all in another. This particular problem has not yet been studied systematically, but ultimately the predispositions of major sectors of the population in *each country* need to be investigated in relation to each type of argument, appeal, source, and communication medium, as well as to each type of topic.

All the predictive attributes discussed so far are referred to as "communication-bound" predispositions. But there also appear to be some much more *general* predispositional attributes associated with susceptibility to many different types of persuasive influence, irrespective of the content of the conclusion or the way it is being promoted. Recent studies indicate that some persons consistently tend to be resistant to *all* types of persuasive communications, whereas other persons consistently tend to be highly persuasible.

In one of our studies, which was designed to investigate general persuasibility among adolescents in high school, we used a procedure consisting of: (1) an initial questionnaire containing fifteen opinion questions; (2) a booklet containing five persuasive communications on widely varying topics; and (3) another booklet presenting a second series of five persuasive communications on exactly the same topics as the first series but taking diametrically opposite positions. After each communication in each of the booklets, the subjects were asked the same opinion questions as in the initial questionnaire.

The communications were intentionally varied with respect to type of appeal and arguments. They were all attributed to an identical type of source: reporters writing on "opinions in the news." The existence of widespread diversity of opinion on the topics was stressed for the subjects, and they were encouraged to express their own views. A "general persuasibility" score was assigned to each subject on the basis of his opinion changes in response to both booklets. The subjects who received the highest scores were the ones who had changed their initial opinion in one direction, in response to the communications in the first booklet, and who then, after exposure to the second booklet, had changed all the way back to their original position or beyond on all the opinion scales.

The results from this procedure showed *consistent individual differences* in persuasibility. Some subjects shifted first in one and then in the other direction following receipt of the conflicting communications in the two booklets, while other subjects did not change their opinions at all. A factor analysis was computed from the intercorrelations between persuasibility subscores on each communication. The results of this study support the hypothesis of a general factor in persuasibility, and they indicate that the predisposition to change opinions is not wholly specific to the topic or subject matter or type of appeal used in the persuasive communications. Similar results, indicating that there are consistent individual differences in readiness to be influenced by persuasive communications, have been found in studies of young children, as well as in other studies of high school students and college students.

We come now to one of the central questions with which research on general persuasibility has been primarily concerned: What are the personality characteristics of the people who are *most resistant* to all forms of persuasion? How do they differ from persons who are *moderately* responsive and from those who are *highly persuasible?*

I shall summarize the research bearing on five hypotheses concerning personality correlates of persuasibility, which are the main ones supported by the evidence that we now have available:

1. Men who openly display overt hostility toward the people they encounter in their daily life are predisposed to remain relatively uninfluenced by any form of persuasion.

The most extreme instances of resistance to persuasion are found among men who display clear-cut paranoid tendencies, or who display overt anti-social behavior of the type encountered in the classical "criminal psychopaths." Among clinically normal men, a high degree of overt aggressiveness also appears to be associated with a low degree of persuasibility. This relationship may be due to an underlying latent need to demonstrate their *power* over others, coupled with strong defense mechanisms against any form of passive behavior. In any case, the strong oppositional tendencies manifested by such personalities make them a poor risk from the standpoint of a communicator who is trying to select strategic individuals toward whom his persuasive efforts are to be directed.

2. Men who display social withdrawal tendencies are predisposed to remain relatively uninfluenced by any form of persuasion.

This hypothesis is not limited to persons with schizoid or narcissistic character disorders but applies also to those clinically "normal" persons who display such characteristics as the following: aloofness, weak and unreliable emotional attachments to love objects, marked preference for seclusive activities, and few affiliations with formal or informal groups within the community.

3. Men who respond with rich imagery and strong empathic responses to symbolic representations tend to be more

persuasible than those whose fantasy responses are relatively constricted.

The relationship between "richness of fantasy" and persuasibility is consonant with the theoretical assumption that a major mediating mechanism in the process of attitude change is the anticipation of rewards and punishments, offered implicitly or explicitly by the communicator as inducements for accepting his conclusions. Men with a rich fantasy life would presumably have greater facility than others in imagining the anticipated consequences conveyed in persuasive communications.

If this hypothesis proves to be a valid generalization for people in all countries, it would have some important implications. For example, one would expect that when new international proposals are made to trade union leaders, government officials, scientists and intellectuals interested in reducing internationl tensions, the individual leaders who are most likely to accept a new point of view concerning means for maintaining world peace will be the ones whose speeches, writings, and other verbal utterances show that they characteristically have rich imagery and are capable of a high degree of empathic response.

4. Men with low self-esteem—as manifested by feelings of personal inadequacy, social inhibitions, and depressive affect —are predisposed to be more readily influenced than others when exposed to any type of persuasive communication.

This propostion applies not only to men who are pathologically depressed but also to a large number of clinically "normal" persons whose emotional conflicts give rise to low self-esteem. Typical indicators of low self-esteem are: feelings of shyness, lack of self-confidence in ability to deal with everyday social situations, high concern about the possibility of being rejected by one's friends, uneasiness at social gatherings, passive compliance to authority, excessive timidity in asserting personal wishes, periodic feelings of sadness, discouragement, and hopelessness.

Men with characteristics of this sort tend to be "passive-dependent" personalities who are generally acquiescent to any source of social pressure. The readiness with which they yield to persuasive communications seems to be a form of social compliance that stems from inability to tolerate anticipated disapproval for deviating from the opinions held by others. Such personalities tend to be indiscriminately influenced by any source of persuasive influence and consequently are likely to show chameleon-like changes in their attitudes. Being strongly motivated to ward off the anticipated disapproval of any communicator, or of any powerful group whose norms are brought to the focus of their attention, such persons are inclined to adopt—at least temporarily—whatever ideas are being promoted. Thus, their excessive compliance can be viewed as a psychological defense that permits them to agree with everyone and to feel that nobody will be displeased.

5. Men with an "other-directed" orientation are predisposed to be more persuasible than those with an "inner-directed" orientation.

This fifth and last hypothesis is closely related to the previous hypothesis. "Other-direction" refers to a value system stressing group conformity and adaptation to the social environment, as against "inner-direction," which involves more emphasis on personal goals and internal standards regulating one's conduct. The positive correlation between other-direction and persuasibility provides the basis for using this personality value system to identify the potential waverers who are most likely to be influenced by an educational or promotional campaign designed to modify any type of belief or attitude. But like those with low self-esteem, their changes in beliefs, expectations, and attitudes are likely to be very short-lived if they are exposed to counterpropaganda.

Thus, when the communicator's goal is to induce long-range actions or persistent attitudes, hyperinfluencible personalities of the sort specified by the last two hypotheses are likely to be only temporarily influenced. On the other hand, when the communicator's goal is merely to induce people to spread information, or to engage in some other form of short-range activity, these hyperinfluencible personalities can be counted on to be among the participants who will augment the success of an educational or promotional campaign.

The reader will notice that all five of the hypotheses refer only to males. This is because in practically all of our studies of personality attributes we have found significant correlations with persuasibility for males *but not for females*. We suspect that this is because of certain sex-typed roles that incline women to take less interest in social, political, and intellectual issues. In this connection, it is important to take account of the fact that all the available evidence supports the popular belief that in impersonal matters girls are more persuasible than boys. Thus, when a series of persuasive communications dealing with political or social issues is given to a large audience, we find more attitude change among females than among males; but, whereas the five hypotheses I have presented enable us to predict *which males* will be most influenced, we have not as yet found any parallel attributes that enable us to predict *which females* will be most influenced. This is one of the problems that awaits further research.

Of course, all the hypotheses I have presented will have to be checked carefully by parallel research investigations with people of all ages and of all educational levels, with samples of different ethnic backgrounds and different nationalities. Perhaps the most serious limitation of the research findings I have described is that practically all the studies were carried out with American high school and college students, or with other limited subgroups within the United States. We cannot be at all sure that the relationships we have found will hold true for people in other parts of the world. Consequently, we hope that research psychologists and social scientists in many other countries will soon begin to check on our findings and to investigate related problems concerning susceptibility to persuasion.

Ultimately, as we learn more about the motivations of those persons who are exceptionally resistant to social influences, we can build up a much more complete picture of "what it takes" to be responsive to new ideas and to be capable of changing one's mind in the face of strong persuasive arguments and appeals. Moreover, by continuing to compare people who are *moderately persuasible* with those who are *most gullible,* we can expect to increase our understanding of "what it takes" to be a *discriminatory* and *flexible* personality—the type of thinking person who can react *selectively* and *rationally* to the persuasive pressures encountered in daily life, accepting some of them and rejecting others, depending upon what is said, how it is said, and who says it.

SUGGESTIONS FOR FURTHER READING

Readers are advised to begin with the following volume, which, in addition to being important for its own sake, also contains a long bibliography on the subject:

JANIS, IRVING L., HOVLAND, CARL I., *et al. Personality and Persuasibility.* New Haven: Yale University Press, 1959.

6 THE SOCIAL EFFECTS OF MASS COMMUNICATION

Joseph T. Klapper

What do the mass media do to people? On the one hand, it is argued that social control is exercised by personal influence, rather than by mass communication, and that the most notable effect of the mass media is to maintain the status quo. On the other hand, a great deal of apprehension has been expressed about the possible influence of television on children, the possible effects of propaganda, and such matters. Dr. Klapper, in the following paper, considers these different points of view, and tries to state some principles to explain the effects of the media.

Born in 1917 in New York, Dr. Klapper graduated from Harvard, earned an M.A. at Chicago and a Ph.D. at Columbia. He began as a teacher of English but soon turned to the study of public opinion and communication. His Ph.D. is in sociology. For a number of years he was associated with the Bureau of Applied Social Research at Columbia, served briefly with the Washington Public Opinion Laboratory, and from 1958 until 1962 was a member of the behavioral research service of the General Electric Company. In 1962, he was appointed Director of Social Research for the Columbia Broadcasting System. He is best known for his volume, The Effects of Mass Communication *(1960).*

The title of this paper is extremely broad. Almost any effect which mass communication might have upon large numbers of people could legitimately be called a social effect, for people make up society, and whatever affects large numbers of people thus inevitably affects society.

We might therefore consider any of a thousand different social effects of mass communication—for example, how mass communication affects people's political opinions and voting behavior, or how it affects its audience's purchases of consumer goods. We might also consider somewhat more abstract topics, such as the ways in which mass communication has changed the social structure as a whole and the relationships of the people within it.

Several of these topics, such as the effect of mass communication on voting behavior, are covered by other papers in this book. Others, such as the effect of mass communication on the social structure, have not been the subject of much objective, scientific research, and discussion of them would necessarily involve considerable conjecture. I like conjecture, but I am writing now as a student and practitioner of communications research, and in this role I should restrict myself to some of the things which objective, scientific research has discovered about the effects of mass communication.

It is difficult to deal with so immense a topic in the limited space available here. Perhaps I might best begin by citing some broad general principles which are, I believe, applicable to the effects of mass communication within a vast number of specific topical areas. I will then illustrate these principles by reference to two such specific areas of effect: first, the effect of mass communication upon the aesthetic and intellectual tastes of its audiences and, second, the question of how these audiences are affected by the crime and violence that is depicted in mass communication. I have selected these two topics because they seem to me matters of social importance as well as popular concern, because none of the other papers in this book discusses them at any length, and because a good deal of information pertinent to these topics has been provided by high-quality communications research. I would like to recall, however, that the principles which I will first develop can be applied to many other types of effect as well. And although we may talk today primarily of levels of public taste and of the effects of crime and violence, I believe that the same principles will be helpful guidelines in considering the probable nature of other social effects of mass communication.

The first point I wish to make is rather obvious, but its implications are often overlooked. I would like to point out that the audience for mass communication consists of people, and that these people live among other people and amid social institutions. Each of these people has been subject and continues to be subject to numerous influences besides mass communication. All but the infants have attended schools and churches and have listened to and conversed with teachers and preachers and with friends and colleagues. They have read books or magazines. All of them, including the infants, have been members of a family group. As a result of these influences, they have developed opinions on a great variety of topics, a set of values, and a set of behavioral tendencies. These predispositions are part of the person, and he carries them with him when he serves as a member of the audience for mass communication. The person who hears a radio address urging him to vote for a particular political candidate probably had some political opinion of his own before he turned on the set. The housewife who casually switches on the radio and hears the announcer state that a classical music program is to follow is probably already aware that she does or does not like classical music. The man who sees a crime play on television almost surely felt, before seeing the play, that a life of crime was or was not his dish.

It is obvious that a single movie or radio or television program is not very likely to change the existing attitudes of audience members, particularly if these attitudes are relatively deep-seated. What is not so obvious is that these attitudes, these predispositions, are at work before and during exposure to mass communications, and that they in fact largely determine the communications to which the individual is exposed, what he remembers of such communications, how he interprets their contents, and the effect which mass communications have upon him.

Communications research has consistently revealed, for example, that people tend in the main to read, watch, or listen to communications which present points of view with which they are themselves in sympathy and tend to avoid communications of a different hue. During pre-election campaigns in the United States, for example, Republicans have been found to listen to more Republican-sponsored speeches than Democratic-sponsored programs, while Democrats do precisely the opposite. Persons who smoke have been found to be less likely to read newspaper articles about smoking and cancer than those who do not smoke. Dozens of other research findings show that people expose themselves to mass communication selectively. They select material which is in accord with their existing views and interests, and they largely avoid material which is not in accord with those views and interests.

Research has also shown that people *remember* material which supports their own point of view much better than they remember material which attacks that point of view. Put another way, retention, as well as exposure, is largely selective.

Finally, and in some senses most importantly, perception, or interpretation, is also selective. By this I mean that people who are exposed to communications with which they are unsympathetic not uncommonly distort the contents so that they end up perceiving the message as though it supported their own point of view. Communications condemning racial discrimination, for example, have been interpreted by prejudiced persons as favoring such discrimination. Persons who smoke cigarettes, to take another example, were found to be not only less likely than non-smokers to read articles about smoking and cancer, but also to be much less likely to become convinced that smoking actually caused cancer.

Now it is obvious that if people tend to expose themselves mainly to mass communications in accord with their existing views and interests and to avoid other material, and if, in addition, they tend to forget such other material as they see, and if, finally, they tend to distort such other material as they remember, then clearly mass communication is not very likely to change their views. It is far, far more likely to support and reinforce their existing views.

There are other factors, besides the selective processes, which tend to render mass communication a more likely agent of reinforcement than of change. One of these is the groups and the norms of groups to which the audience member belongs. Another is the workings of interpersonal influence. A third involves the economic aspects of mass media in free enterprise societies. Limitations of space do not permit me to discuss these factors here, but those who are sufficiently interested in this topic will find them all discussed in the literature of communication research.

It will of course be understood that, again because of space limitations, I am writing in terms of general tendencies, and that I cannot here discuss all the exceptions to these general tendencies. I can only say that there are exceptions and that these, too, are discussed in the literature. But the exceptions are, at least in my opinion, precisely that—exceptions. And I have in fact gone so far as to assert, in some writings of my own, and on the basis of the findings of numerous studies performed by numerous people, that the typical effect of mass communication is reinforcive. I have also stated, as I have tried to show in this paper, that this tendency derives from the fact that mass communication seldom works directly upon its audience. The audience members do not present themselves to the radio or the television set or the newspaper in a state of psychological nudity; they are, instead, clothed and protected by existing predispositions, by the selective processes, and by other factors. I have proposed that these factors serve to mediate the effect of mass communication, and that it is because of this mediation that mass communication usually serves as an agent of reinforcement.

Now this does not mean that mass communication can *never* produce changes in the ideas or the tastes or the values or the behavior of its audience. In the first place, as I have already mentioned, the factors which promote reinforcive effects do not function with 100 per cent efficiency. In the second place, and more importantly, the very same factors sometimes maximize the likelihood of mass communications serving in the interest of change. This process occurs when the audience member is *predisposed* toward change. For example, a person may, for one reason or another, find his previous beliefs, his previous attitudes, and his accustomed mode of behavior to be no longer psychologically satisfying. He might, for example, become disillusioned with his political party, or his church, or—on another level—he might become bored with the kind of music to which he ordinarily listens. Such a person is likely to seek new faiths, or to experiment with new kinds of music. He has become, as it were, *predisposed to change*. And just as his previous loyalties protected him from mass communications which were out of accord with those loyalties, so his new predispositions will make him susceptible to the influence of those same communications from which he was previously effectively guarded.

Let us now pause for a moment and look back over the way we have come. I have cited what I believe to be three basic principles about the effects of mass communication. I have stated, first, that the influence of mass communication is mediated by such factors as predispositions, selective processes, group memberships, and the like. I have proposed, secondly, that these factors usually render mass communication an agent of reinforcement. Finally, I have said that these very same factors may under some conditions make mass communication an agent of change. All of this has been said in a rather abstract context. Let us now see how these principles apply in reference to such specific topics as the effect of mass communication on levels of public taste, and the effect of depictions of crime and violence.

We would all agree, I believe, that a great proportion of the material on the mass media is on a rather low aesthetic and intellectual level. The media do, of course, provide classical music, readings and dramatizations of great books, public affairs programs, and other high-level material. But the lesser material greatly predominates. And we are all familiar with the frequently expressed fear that this heavy diet of light fare will debase or has already debased the aesthetic and intellectual tastes of society as a whole. What has communications research discovered in reference to this matter?

Communications research long ago established that the principle of selective exposure held in reference to matters of taste. Persons who habitually read good books were found to listen to good radio programs, and persons who read light books or no books were found to listen to light radio programs. Recent research has indicated that children and young people who like light fiction will tend to seek light entertainment at the television set, and that people who read books on public affairs will find and witness television discussions on public affairs.

Increasing the amount of high-level material on the air has been found to serve very little purpose. There is already a good deal of fine material available on radio and television. Those who like it find it. Those who don't like it turn to other programs which, at least in this country, are almost always available. In short, and in accord with the basic principles I previously mentioned, mass communication generally serves to feed and to reinforce its audience's existing tastes, rather than to debase or to improve them.

But this is by no means to say that mass media are never involved in changing the tastes of their audience. Our third principle, it will be recalled, states that mass communication will change people if they are already predisposed to change. Let me give an example of this principle at work in reference to levels of taste.

Some years ago a student of communication research made a study of persons who listened to certain serious music programs on the radio. He found that the overwhelming majority had long been lovers of serious music, although some of them, for various reasons, had been unable to hear as much of it as they would have liked until radio made it so easily accessible. About 15 per cent of the group, however, considered that the radio had initiated their liking for classical music. But—and here is the essential point—closer analysis revealed that most of these people were predisposed to develop a liking for such music before they began listening to the programs. Some of them, for example, wanted to emulate friends who were serious music lovers. Others had attained a social or occupational status such that they felt they *ought* to be interested in serious music. With these predispositions, they found or sought out the serious music programs, and grew to like that kind of music. Their tastes had indeed been changed by the programs, but they had come to the programs already predisposed to change their tastes. Mass media had simply provided the means for the change.

Findings of this sort inevitably inspire the question of whether it would be possible deliberately to create in people predispositions to enlarge their intellectual and aesthetic horizons, and so nurture a widespread rise in levels of taste. Such hypothetical developments are somewhat beyond the scope of this paper, but I would venture the guess that such a development would be possible if it were sufficiently carefully planned and executed. Children seem to me particularly good subjects for such an attempt, since children are naturally "changers." As they grow older, they naturally change in various physical and psychological ways. Their habits of media usage change too, if only in the sense that they progress from material designed for children to material designed for adults. The problem is, then, to so predispose them that they advance not merely to material designed for adults but continue on to progressively better adult material.

I cannot here go into the findings of research pertinent to this problem, but I will say, by way of summary, that this research indicates that even among children mass media do not so much determine levels of taste, but are, rather, used by the child in accordance with his already existing tastes. These tastes appear to be a product of such extra-media factors as the tastes of the child's parents and peer group members, the nature of his relationship with these people, and the child's own level of intelligence and degree of emotional adjustment. Insofar as these conditions are manipulable by parents, by schools, and by social programs, it would appear possible to develop predispositions for high-quality media material, which predispositions could then be served and reinforced by mass media themselves. I would point out, however, that in such a process the media would be functioning in their usual adjunctive manner. They would not be serving, in and of themselves, to elevate standards of public taste. They would rather be serving to supply a channel of change for which their audience had already been predisposed.

Let us now turn to the question of the effect of crime and violence in mass communication. Everyone will agree, I think, that depictions of crime and violence abound in the media. And we are all familiar with the widely expressed fear that these portrayals will adversely affect the values and behavior of the media audiences, possibly to the point of individuals actually committing criminal violence. Communications research, for all its attention to this matter, has not yet provided wholly definitive conclusions. The accumulating findings, however, seem to indicate that the same old principles apply.

A large number of studies have compared children who are heavy consumers of crime and violence material with children who consume little or none of it. Many of these studies have found no differences between the two groups: heavy users were found, for example, to be no more likely than light users or non-users to engage in delinquent behavior, or to be absent from school, or to achieve less in school. Other studies, which inquired more deeply into the psychological characteristics of heavy and light consumers, have found differences between the two groups. Heavy users have been found, for example, in one or another study, to be more likely to have problems relative to their relationships with their families and friends, to place blame for difficulties on others rather than themselves, to be somewhat more aggressive, and to have somewhat lower I.Q.'s. Children who did not have satisfactory relations with their peers were found in one study not only to be particularly drawn to such material but also to employ it as a stimulant for asocial fantasies. The children with good peer relations, on the other hand, employed the same material as a basis for group games.

Let us now draw some implications from these findings. First of all, since both delinquent and non-delinquent children are found among heavy users of crime and violence material, we may assume that the material is not in and of itself a prime cause of delinquent tendencies. Secondly, such differences as have been found between heavy and light users consist of personality and emotional factors which seem unlikely to have been the *product* of exposure to the media. Finally, the uses to which the material is put appears to be dependent upon these same personality factors. Here again, then, our old principles seem to be at work: children appear to interpret and react to such material in accordance with their existing needs and values, and the material thus serves to reinforce their existing attitudes, regardless of whether these existing attitudes are socially wholesome or socially unwholesome. The media, as usual, seem to be not a prime determinant of behavioral tendencies but rather a reinforcing agent for such tendencies.

Our basic principles would, however, lead us to expect that the media might play a role in changing the values and behavioral tendencies of audience members who were, for one reason or another, predisposed to change. Unfortunately, I know of no research which throws any light on this topic in relation to tendencies to criminal or violent behavior. Several such studies are now in progress, but none has yet reached the reporting stage.*

Here again, as in the case of the discussion of levels of taste, one must inevitably wonder how the undesirable effects might be minimized. And here again the nature of existing research findings suggests that the road cannot involve the media alone. Remedies, if they can be defined at all, seem likely to involve the family, the schools, and all of those forces which create the values and the personality which the child, or the adult, brings to the media experience.

* *Editor's note:* Dr. Klapper's expectations have recently been borne out by a group of studies which have shown that children with aggressive tendencies are more likely than less aggressive children to imitate violent behavior they see in films and television.

So much, then, for the effects of depictions of crime and violence, and so much for the effects of mass communication on levels of public taste. In the brief space available to me I have not, of course, presented all aspects of the story, but I have tried to present a general picture which I believe is valid for other types of social effects as well. Research strongly suggests, for example, that the media do not engender passive orientations toward life, nor do they stimulate passively oriented persons to activity. They seem to provide a passive activity for the already passive and to stimulate new interests among persons who are intellectually curious, but they rarely change one type of person to another. In general, mass communication reinforces the existing attitudes, tastes, predispositions, and behavioral tendencies of its audience members, including tendencies toward change. Rarely, if ever, does it serve alone to create metamorphoses.

This is, of course, not to say that mass communication is either impotent or harmless. Its reinforcement effect is potent and socially important, and it reinforces, with fine disinterest, both socially desirable and socially undesirable predispositions. Which are desirable and which undesirable is, of course, often a matter of opinion. I have tried to show, however, that reducing such effects as may be considered undesirable, or increasing those which are considered desirable, is not likely to be achieved merely by modifying the content of mass communication. Mass communication will reinforce the tendencies which its audience possesses. Its social effects will therefore depend primarily on how the society as a whole —and in particular such institutions as the family, schools, and churches—fashions the audience members whom mass communication serves.

I would like to mention briefly a few other points. The first and most important of these is long-range effects. I have concentrated largely on short-term effects in this paper, for these are the effects with which research has been concerned. Next to nothing is known as yet regarding the social effects of mass communication over periods of, let us say, two or three decades. A second topic I have omitted is the power of mass communication in creating opinions on new issues—that is, issues on which its audience has no predispositions to reinforce. In related vein, I must mention that the media are quite effective in changing attitudes to which audience members are not particularly committed, a fact which explains much of the media's effectiveness in advertising. All of these topics are discussed in the appended list of books and articles.

SUGGESTIONS FOR FURTHER READING

1 HIMMELWEIT, HILDE T., OPPENHEIM, A. N., and VINCE, PAMELA. *Television and the Child.* Published for the Nuffield Foundation. London and New York: Oxford University Press, 1958.

2 KATZ, ELIHU, and LAZARSFELD, PAUL F. *Personal Influence: The Part Played by People in the Flow of Mass Communications.* Glencoe, Ill.: The Free Press, 1955.

3 KLAPPER, JOSEPH T. *The Effects of Mass Communication.* Glencoe, Ill.: The Free Press, 1960.

4 LAZARSFELD, PAUL F., BERELSON, BERNARD, and GAUDET, HAZEL. *The People's Choice.* New York: Columbia University Press, 1948.

5 LAZARSFELD, PAUL F., and STANTON, FRANK N., eds. *Communications Research, 1948–1949.* New York: Harper & Brothers, 1949.

6 ———. *Radio Research, 1941.* New York: Duell, Sloan and Pearce, 1941.

7 SCHRAMM, WILBUR, ed. *The Process and Effects of Mass Communication.* Urbana, Ill.: University of Illinois Press, 1954.

8 SCHRAMM, WILBUR, LYLE, JACK, and PARKER, EDWIN B. *Television in the Lives of our Children.* Stanford: Stanford University Press, 1961.

7 THE DIFFUSION OF NEW IDEAS AND PRACTICES

Elihu Katz

How does a new product or practice come into use? Obviously, this is at least partly a communication problem, as the merchandisers recognize when they promote a new product, and as the agricultural extension service recognizes when it sends out county agents and supports them with broadcasts, pamphlets, visual aids, and newspaper publicity. Dr. Katz analyzes the processes by which knowledge of new ideas and practices reaches the public, and the communication chains by which individual users are persuaded to adopt them.

Dr. Katz is one of the several alumni of the Bureau of Applied Social Research who appear in this series. Born in Brooklyn in 1926, he earned three degrees at Columbia University, his Ph.D. in sociology coming in 1956. He taught for two years in Israel, then joined the faculty at the University of Chicago, where he is an associate professor of sociology. He is perhaps best known for the book he wrote jointly with Paul Lazarsfeld, Personal Influence *(1956). In recent years, he has been working on diffusion and adoption, and is the author of a book jointly with James S. Coleman and Herbert Menzel on the patterns of adoption of new drugs by physicians.*

Rapid social and technical change is a hallmark of the American society and in recent years the pace of change has quickened all over the world. Accounting for the dynamics of change is surely one of the generic problems of social science, although the empirical study of change processes is still very young. One of the processes that has received systematic attention is that of diffusion, which I will define as the process of spread of a given new idea or practice, over time, via specifiable channels, through a social structure such as a neighborhood, a factory, or a tribe.

A number of social science disciplines have confronted the problem of diffusion, though in rather different ways.

77

Modern anthropology, for example, virtually was born out of the controversy between those who conceived of evolutionary processes as the major dynamic of change and those, called "diffusionists," who were trying to reconstruct history in terms of the itinerary of ideas or artifacts as they moved through time and space. Though the debate has long since been abandoned, social anthropologists today are giving much of their attention to the study of the changes that result from the contact of cultures, including the contacts implicit in programs of technical assistance to underdeveloped areas.

Sociology, too, has had a long-standing interest in diffusion. Apart from the important speculative writings about fads, fashions, public opinion formation, and the like, the 1930's saw a number of studies which sought to trace the spread of social innovations, such as cooperative credit societies, hobbies, political movements, and the like, as a function of rural-urban contacts, state and county boundaries, patterns of shopping and transportation, and so on. At about the same time, rural sociologists began to study the diffusion of farming information and farm practices and, to date, have completed almost two hundred studies in this area. In the 1930's, too, with the advent of radio, research on mass communication began.

Closely allied to the sociologists, social psychologists have also been concerned with problems of the diffusion of innovation but have focused rather more on the factors associated with the willingness to accept change. It is true that studies of rumor, for example, have emphasized the patterns of message distortion in word-of-mouth communication or, to choose another example, the role of friendship relations as channels of interpersonal communication. But much more characteristic of social psychology have been studies such as those of Kurt Lewin and his associates, which were concerned with the relative effectiveness of methods of group discussion (compared with lectures or clinical interviews) in gaining acceptance for changes in food habits, child care, technical improvements in a factory, and the like.

Applied researchers in various fields have also contributed. Educational researchers have traced the spread of the kindergarten and other educational innovations, emphasizing the influence of urban centers, of the relations between the school superintendents and the community, and so forth. Marketing research has, of course, been concerned with the diffusion of new products (though it is puzzling to note how little work has been done on diffusion that is of theoretical interest) and researchers in the field of public health have given attention to the acceptance of new health practices, in addition to the traditional· concern with the diffusion of disease which, of course, is a related phenomenon.

Despite these shared concerns, however, these various compartments of social research have shown very little awareness of each other. Only in the last few years have there been signs of a growing interchange and some movement toward a generic approach to the study of diffusion.

From the vantage point of sociology, the locus of present-day concern with diffusion processes is to be found among the rural sociologists on the one hand and, on the other hand, among several groups with a shared background of interest in mass communication and processes of public opinion formation.

In what follows, I should like to take mass communication research as a starting point—for it is my own starting point as well—and to trace the development of theory and research on the diffusion of innovation. I will attempt to do this by discussing a specific sequence of empirical studies. Then, I should like to go on to indicate the point at which contact was made between the tradition of research on mass communication and various other traditions of research on diffusion, but particularly with rural sociology and anthropology. In doing so, I shall also attempt to illustrate the characteristic emphases of these several traditions, with the thought that each can contribute, ultimately, to the formulation of a broadly applicable design for research on the diffusion of new ideas and practices.

Much of the history of mass communication research can be characterized as the study of "campaigns"—that is, attempts, in the very short run, to change opinions, attitudes, and actions. The model in the minds of early researchers seems to have consisted of: (1) the all-powerful media, able to impress ideas on defenseless minds; and (2) the atomized mass audience, connected to the mass media but not to each other.

Empirical research rapidly dispelled this simple set of images and proved how difficult it is to "convert" people by means of mass media alone. Dozens of studies, particularly in the field of public affairs, have failed to find any appreciable change in opinions, attitudes, or actions resulting from a mass media campaign. The audience, at least in a democratic society, exposes itself to what it wants to hear; even a captive audience engages in a kind of motivated missing-of-the-point when it finds its prejudgments under fire. So, the mass media were found to be far less potent than had been supposed (though popular writings still frequently equate the mass media with "brainwashing").

A major turning point in the conceptualization of the process of mass communication is to be found in the study by Lazarsfeld, Berelson, and Gaudet of decision-making in the 1940 presidential campaign. Using a panel technique of repeated interviews with the same sample of respondents, they tried to examine the actual process of deciding to vote one way rather than another. Reconstructing the respondents' decisions, they found that the mass media—radio speeches, newspaper editorials, and so on—had far less effect on votes than they had hypothesized. First of all, very few people changed their voting intentions during the campaign. Among those who did, however, the major source of influence (despite the original image of the atomized audience) appeared to be other people: family, friends, co-workers.

Taking one additional step, the authors located and interviewed the people whom they called "influentials" or "opinion leaders"—those who had exerted influence on someone else's vote—and compared them with the others. They found the influentials very similar to the non-influentials in social-class status, education, age, and the like. The groups differed, however, in their communications behavior: the opinion leaders were considerably more exposed to mass media than were the others. Together, these findings led to the hypothesis of the "two-step flow of communication," which proposed that influence moves from the media to opinion leaders and from such everyday influentials to their intimate associates.

Several studies built directly on this idea. One study, in a New Jersey town, located opinion leaders by asking a representative sample of the population to name those to whom they would turn if they wanted information or advice on a variety of subjects. Another study, among housewives in a Midwestern city, painstakingly reconstructed all of the sources of influence that went into the making of very specific, very recent decisions (to try a new brand of coffee, to attend a particular movie, to make or buy more fashionable clothing, to vote one way rather than another in a referendum). This was the study in which a technique of "snowball sampling" was attempted: whenever a respondent implicated another person as having been influential in the making of her decision, the interviewer was instructed to obtain the name and address of the designated influential and to interview her (or him) as well. The object was to obtain a picture of the two parties to the influence-exchange and to learn something of the situation and the dynamics of influence. A third study, dealing with patterns of communication in the Middle East, explored the workings of systems of oral communication in traditional villages on the threshold of change; particular attention was given to the conditions making for stability and instability of opinion leadership in situations of incipient change.

These studies, all products of the Bureau of Applied Social Research at Columbia, gave considerable support to the idea that interpersonal communication networks mediate between the mass media and their individual targets. In general, they found that: (1) the influence of other people on specific decisions tends to be more frequent, and certainly more effective, than the influence of the media; (2) influentials are close associates of the people whom they influence and, hence, tend to share the same social status characteristics—it is almost as rare to find somebody of higher status directly influencing somebody of lower status as vice versa; (3) intimate associates tend to hold opinions and attitudes in common and are reluctant to depart unilaterally from the group consensus even when a mass media appeal seems attractive; (4) there is specialization in opinion leadership—a woman who is influential in marketing is unlikely to be influential in, say, fashion; (5) while influence tends to move from persons who are more interested in a given subject to persons who are less interested, the latter have to be sufficiently interested to be susceptible—there are no leaders unless there are followers, and followership requires interest; (6) opinion leaders tend to be more exposed to the mass media, particularly to the media most relevant to their spheres of influence.

Along with the realization of the inadequacy of the model of the atomized mass society which guided early communication research, there came an awareness of the need for theoretical and methodological reformulations. One such development was the rapprochement between mass communication research and small-group research. The idea that these two compartments of society and of social research might have something to contribute to each other would have been totally discredited two decades ago. Yet the finding that interpersonal influence plays such an important role in the process of mass communication suggested that the study of small-group dynamics might well advance the study of communication. As a result, concepts such as Kurt Lewin's "gatekeeper" (analogous, in a number of ways, to the concept of "opinion leader") were found to be immediately relevant.

A second development was methodological. As a result of the growing interest in interpersonal influence, it became clear that the sample survey—the major tool for the study of public opinion and communication—was not altogether adequate. So long as individuals were conceived as unconnected atoms responding individualistically to sources of remote control such as the mass media, or law, a representative sample of individuals as few and as "far apart" as possible was perfectly appropriate. To study interpersonal influence, however, one needs individuals who are "close together," so to speak. It is this requirement that led to some of the methodological innovations mentioned above, such as the "snowball" sample.

A third development was a concern for the incorporation of time as a variable. The early model of mass communication assumed a kind of stimulus–response process such that people immediately reacted (or didn't react) to an influence attempt. Given the revision of this model implicit in the foregoing, it became clear that influence or innovation spreads gradually through society via various combinations of mass media and interpersonal networks. To trace the spread of influence, therefore, it was obviously necessary to introduce the element of time. The introduction of time was the key to the reorientation of mass communication research from the study of "campaigns" to the study of diffusion.

These three developments are well represented in a study of the diffusion of new drugs among physicians. Doctors in four Midwestern communities were selected for study and, instead of drawing the usual kind of sample, a decision was made to include the total population of physicians in these communities. Thus, along with an elaborate interview schedule which inquired into each doctor's background, attitudes, styles of diagnosis, experience with specific new drugs, and so on, there was developed (1) a sociometric questionnaire, by means of which it became possible to construct the patterns of friendship, informal consultation, and advice-seeking among all respondents; and (2) an audit of prescriptions in all local pharmacies, by means of which it became possible to date each physician's first use of the particular new drug which was chosen for study.

Of all the factors found relevant to speed of adoption of the new drug, a doctor's integration in the networks of interpersonal communication was about the most important. The more often a physician was named by colleagues as a friend, or as somebody who shared ideas about drugs and therapy, or as somebody worth consulting for information or advice about medicine, the earlier the introduction of the new drug into his practice. Although the drug did not approach total acceptance until some two years after having become available, physicians who maintained a large number of professional contacts tried it months before their more "isolated" colleagues. Moreover, the early-adopting physicians were considerably more likely to be in touch with relevant sources of influence outside their community as well as inside it: they were more likely to read extensively in their professional journals, more likely to be in touch with an out-of-town research center, more likely to attend out-of-town specialty meetings. In short, just as the "opinion leaders" in the earlier studies were found to be more attuned to relevant messages from the mass media, the innovating physicians were more likely to be connected with relevant information sources in the world outside their immediate professional communities for which, in turn, they serve as relayers.

The drug study also began to explore the shape of diffusion curves and, in constructing tentative models reflecting "social" and "individual" processes of diffusion, found that the diffusion curve for sociometrically integrated doctors fit a "social" model in which it is assumed that the earliest users of the innovation influence their associates who, in turn, influence *their* associates, and so on. The diffusion curve for isolated doctors, on the other hand, fit an "individualistic" model in which it is assumed that some constant source of influence is equally effective at each point in time.

It is interesting that another group of researchers—whose intellectual history also reflects a continuing interest in public opinion—became interested in the mathematical properties of diffusion curves at about the same time. This is the group which centers around Stuart C. Dodd at the University of Washington, and which has carried out a large number of recent studies on the diffusion of information via word of mouth. The starting point for most of the studies in this series involves dropping leaflets from the air over a town and then, by means of interviews, reconstructing the itinerary, speed, and effectiveness of diffusion of the message as a function of variations in the leaflet–population ratio, the population density together with other measures of social structure, and the type of reward offered for learning and passing on the message. For example, "Diffusion [of information] grew absolutely but decreased per capita in a somewhat harmonic curve as community size went from 1,000 to over 300,000, i.e., small towns seem to gossip more." Or, to choose an example from the realm of social roles, children were more effective than adults in circulating the message.

It was not until after the completion of the drug study that students of diffusion who derived their interest from mass communications "discovered" that other traditions of social research had been sharing the same problem, namely, the problem of how social systems incorporate a change which originates "outside" the system. The two major traditions were rural sociology and anthropology.

There is good reason to believe that the barrier that existed between mass communication researchers, on the one hand, and rural sociologists and anthropologists, on the other, is a product of the contrasting pictures of society in the minds of each. If the model underlying early research in mass communication is that of an atomized mass society, the rural sociologist and the anthropologist, obviously, were working at the other end of the continuum. Their model was one of the traditional rural society, based primarily on personal relations. It is for this reason that the student of mass communication had to "rediscover" interpersonal relations, while the student of new farm practices took them for granted. But now it is clear that the campaign of a county agent or a manufacturer of farm implements to gain acceptance for a new farm practice is not different, in any theoretical sense, from the campaign of a drug company to gain acceptance for a new drug.

Indeed, a comparison of the drug study with a study—completed fifteen years earlier—on the diffusion of hybrid-seed corn in two Iowa communities reveals a number of striking similarities. Although it took ten years for the new seed to gain the same level of acceptance as the new drug gained in two, there are many parallels in the underlying processes.

Like the drug study, the seed study took account of time (by asking the farmer to recall the season during which he first had tried hybrid-corn seed). It took account of social relations (by studying patterns of neighboring, and investigating the role of interpersonal influence in decisions to adopt the innovation). The seed study concluded that the early adopters influenced the later adopters, and that the early adopters were much influenced by salesmen and farm bulletins and more frequent trips to the city—again, the "two-step flow of communication." Like the drug study, it found that the media and commercial sources bring first news of an innovation, but that colleagues, friends, and trusted professional sources are required to "legitimatize" decisions.

A major difference between the two studies is the finding, in the drug study, that social integration is positively related to time of adoption, whereas the seed study finds that extent of informal neighboring and other measures of social relations are not predictive of time of adoption. Another rural sociological study has attempted a reconciliation of these findings: If "progressive" counties are compared with "conservative" counties (measured by proportion of recommended farm practices adopted over a period of years), it is found that the influential members of "progressive" communities will have adopted a greater number of recommended practices than the average member, whereas the influential members of the "conservative" communities will have adopted no more than average. In other words, where a group norm affirms change, as it presumably does in a medical community, there is reason to expect integration to be positively related to innovation. In a traditional community, however— the two Iowa communities were considered so by the authors of the seed study—it may well be the less integrated members who innovate.

This inverse relationship between integration and innovation is frequently found in anthropological studies of reactions to technical assistance campaigns. In a study of a concerted effort to gain acceptance for the practice of boiling drinking water in a Peruvian town, for example, it was found that the most marginal and disaffected members of the community were most willing to try. This has been the experience (and the doom) of many such campaigns, for if low-status individuals become "demonstration farmers," a campaign to gain acceptance for change has thereby lost much of its chance for success.

A review of the studies of diffusion in the fields of rural sociology and anthropology reveals characteristic emphases which have been relatively neglected in the diffusion studies reported so far; and, by the same token, some of the emphases of the latter are absent in the rural and anthropological reports.

Both rural sociologists and anthropologists, first of all, have done more *comparative* studies. For example, a rural sociological study finds that cohesive residential areas ("neighborhoods") tend to obtain farming information primarily from interpersonal sources and only secondarily from the mass media, while matched "non-neighborhoods" rely prima-

87

rily on the mass media. While such studies tend to find that the more cohesive, more traditional communities are slower to accept change, this is by no means always the case. In fact, as the variable relationship between individual integration and innovation suggests, it is the value and the meaning that the group attributes to change which provide the best explanation. Thus, according to a recent comparative-anthropological account, the more traditional, more strongly centralized kingdom of Ruanda has been considerably more receptive to European influences (in the school system, religion, economic and political institutions, language) than the less traditional, more loosely organized people of Urundi.

The rural sociologists have been particularly interested in a theory of stages of the diffusion process as reflected in the over-all movement of an innovation through a community and in the process of individual decision-making. They have also been interested in the functional appropriateness of different media for the various stages. Reviewing a number of studies, Herbert Lionberger suggests that their results appear to fit an "S" curve implying: (1) an early stage, in which a small number of "cosmopolitan" adopters try the innovation, relying for their information on agricultural-station bulletins, on mass media, and on other highly competent farmers; (2) a middle stage, in which the majority adopts rather quickly, influenced primarily by the innovators; and (3) a late stage, in which older, more isolated, more security-minded farmers finally capitulate. Decision-making stages appear to go on within the individual as well: mass media or commercial sources bring the earliest information and constitute the "awareness" stage; at the "interest" stage a variety of sources are used to get further information; at the "evaluation" stage—when the potential adopter is considering the applicability of an innovation for himself—respected fellow farmers constitute the most important source of influence; there follow the "trial" and "adoption" stages. Studies have established the empirical validity of those hypothesized stages, the extent to which different media are called upon at each stage, and the consequences of using the "wrong" sources for a given stage.

Unique to the anthropological tradition of research on diffusion is a focus on intergroup relations rather than intragroup relations. Thus the structure of social relations characterizing "donor" and "borrower" cultures, or the relationship between colonial administrators and their wards, have been examined for their effect on the probability that change will diffuse. Within a group, anthropologists have analyzed the influence of clique structures, vested interests, and patterns of cleavage for their impact on the itinerary of social and technical change.

Another basic anthropological emphasis is the one on the role of *culture*, of pattern-of-values, acting as a sort of filter which permits only certain changes to enter and substantially modifies those that get through. Thus, in a study of the response of American Indians to Christianity, it was found that patrilineally organized tribes were more receptive than matrilineally organized tribes by virtue of the patrilineal symbolism of Christianity. Or the concept of boiling drinking water has been resisted where theories of "hot" and "cold" pervade ideologies of food and health. Given this theoretical focus on "compatibility" between an innovation and a pattern of culture, it is easy to explain how ostensibly incompatible innovations which are accepted may, in fact, be transformed in the process. Thus the temperate, self-disciplined Zuni accepted the peyote flower but did not use it in the orgiastic manner of their neighbors. The notion of compatibility has been applied outside of anthropology primarily in marketing and motivation research: television, for example, was adopted early by families who were relatively more oriented to the present than to the past or future.

Again, anthropologists have tended to treat the community or the tribe as the *unit of adoption;* sociologists have tended to treat the individual as such. Some recent theorizing on this subject, however, suggests that this matter deserves more attention. First of all, it seems that different kinds of innovations "require" different kinds of adopting units: hybrid corn or television, for example, are aimed at individual adopters, but other innovations such as the telephone, or the sustained use of contraceptives, or the introduction of the peyote cult among American Indians, "require" cooperating groups for implementation. The popular song of some years ago, "It Takes Two to Tango," illustrates the point. Still a third type of innovation also requires a group decision but, unlike the telephone or the tango, leaves no room at all for non-adopters. The fluoridation of a community's water supply is an example of this kind of innovation.

It is also interesting to observe that certain societies show a decided preference for one or another unit of adoption. This is well illustrated in the recent explanation offered by the United States Surgeon General as to why the United States was slow to certify the new oral vaccine for use against infantile paralysis, even though the Soviet Union had long since done so and with apparent success. Given that the decision to be inoculated is an individual matter, he explained, there is some danger in the case of this particular vaccine that those who are inoculated may infect those who are not. In the Soviet Union, he implied, whole communities are inoculated simultaneously. In our terms, this is an innovation which "requires" a group of adopters to act simultaneously and thus is more likely to be adopted earlier by a society in which the group, rather than the individual, is the preferred unit of adoption.

Such examples might be multiplied, of course, but perhaps the point is clear. There exist a large number of studies of diffusion organized within a variety of different research traditions. Each tradition has its characteristic approach, emphasizing certain concepts and methods rather than others, although there are also basic similarities.

Until very recently, there was virtually no awareness

shown by research workers in the various traditions that colleagues in other traditions were concerned with essentially the same set of problems. I have tried to sketch the way in which mutual awareness has begun to come about, particularly among students of mass communication, rural sociology, and anthropology. To a lesser extent, it is also true of marketing and public health. I have also tried to show some of the parallels in research methods and results, as well as to point out some of the ways in which the characteristic emphases of the several traditions tend to complement each other.

It seems to me that there is good reason to expect interesting collaboration among students of diffusion and the emergence of a kind of generic research design for increasing our understanding of one of the central processes of social and technical change.

SUGGESTIONS FOR FURTHER READING

1 ALBERT, ETHEL M. "Socio-Political Organization and Receptivity to Change: Some Differences Between Ruanda and Urundi," *Southwestern Journal of Anthropology*, 1960, *16*, pp. 46–74.

2 BEAL, GEORGE M., BOHLEN, JOE M., and ROGERS, EVERETT M. "Validity of the Concept of Stages in the Adoption Process," *Rural Sociology*, 1957, *22*, pp. 166–168.

3 COLEMAN, JAMES, MENZEL, HERBERT, and KATZ, ELIHU. "Social Processes in Physicians' Adoption of a New Drug," *Journal of Chronic Diseases*, 1958, *8*, pp. 1–19.

4 COLEMAN, JAMES S., KATZ, ELIHU, and MENZEL, HERBERT. *Doctors and New Drugs* (in press).

5 COPP, JAMES H., SILL, MAURICE L., and BROWN, EMORY J. "The Function of Information Sources in the Farm Practice Adoption Process," *Rural Sociology*, 1958, *23*, pp. 146–157.

6 DeFleur, Melvin L., and Larsen, Otto N. *The Flow of Information*. New York: Harper & Brothers, 1958.

7 Dodd, Stuart C. "Formulas for Spreading Opinions," *Public Opinion Quarterly*, 1958–59, 22, pp. 537–554.

8 Freidson, Eliot. "Communication Research and the Concept of Mass," in Wilbur Schramm, ed., *The Process and Effects of Mass Communication*, Urbana, Ill.: University of Illinois Press, 1954.

9 Graham, Saxon. "Cultural Compatibility in the Adoption of Television," *Social Forces*. 1954, 33, pp. 166–170.

10 Hawley, Florence. "The Role of Pueblo Social Organization in the Dissemination of Catholicism," *American Anthropologist*, 1946, 48, pp. 407–415.

11 Katz, Elihu. "Communication Research and the Image of Society: the Convergence of Two Traditions," *American Journal of Sociology*, 1960, 65, pp. 435–440.

12 ———. "The Social Itinerary of Technical Change: Two Studies on the Diffusion of Innovation," *Human Organization*, 1961, 20, pp. 70–82.

13 ———. "The Two-step Flow of Communication: An Up-to Date Report on an Hypothesis," *Public Opinion Quarterly*, 1957, 21, pp. 61–78.

14 Katz, Elihu, and Lazarsfeld, Paul F. *Personal Influence: The Part Played by People in the Flow of Mass Communications*. Glencoe, Ill.: The Free Press, 1955.

15 Klapper, Joseph T. "What We Know About the Effect of Mass Communication: The Brink of Hope," *Public Opinion Quarterly*, 1957, 21, pp. 453–474.

16 Kroeber, A. L. "Diffusionism," in Edwin B. A. Seligman and Alvin Johnson, eds., *Encyclopedia of the Social Sciences*, New York: Macmillan, 1930.

17 Lazarsfeld, Paul F., Berelson, Bernard, and Gaudet, Hazel. *The People's Choice*. New York: Columbia University Press, 1948.

18 Lewin, Kurt. "Group Decision and Social Change," in Eleanor Maccoby, Theodore Newcomb, and Eugene Hartley, eds., *Readings in Social Psychology*. New York: Henry Holt and Co., 1958.

19 LIONBERGER, HERBERT F. *Adoption of New Ideas and Practices.* Ames, Iowa: Iowa State University Press, 1960.

20 MARSH, C. PAUL, and COLEMAN, LEE. "Group Influences and Agricultural Innovations: Some Tentative Findings and Hypotheses," *American Journal of Sociology,* 1956, *61.*

21 MERTON, ROBERT K. "Patterns of Influence: A Study of Interpersonal Influence and Communications Behavior in a Local Community," in Paul F. Lazarsfeld and Frank Stanton, eds., *Communications Research 1948–49.* New York: Harper & Brothers, 1949.

22 MORT, PAUL R., and CORNELL, FRANCES G. *American Schools in Transition.* New York: Teachers College, Columbia University, 1941.

23 North Central Rural Sociological Subcommittee. *Social Factors in the Adoption of Farm Practices.* Ames, Iowa: Iowa State College, 1959.

24. PAUL, BENJAMIN D., ed., *Health, Culture, and Community.* New York: Russell Sage Foundation, 1955.

25 ROGERS, EVERETT M., and BEAL, GEORGE M. "The Importance of Personal Influence in the Adoption of Technological Change," *Social Forces,* 1958, *36,* pp. 329–340.

26 RYAN, BRYCE, and GROSS, NEAL. "The Diffusion of Hybrid Seed Corn in Two Iowa Communities," *Rural Sociology,* 1943, *8,* pp. 15–24.

27 WELLIN, EDWARD M. "Water Boiling in a Peruvian Town," in Benjamin Paul, ed., *Health, Culture, and Community.* New York: Russell Sage Foundation, 1955.

28 WILKENING, EUGENE. "Roles of Communicating Agents in Technological Change in Agriculture," *Social Forces,* 1956, *34,* p. 361.

8 MASS MEDIA AND PERSONAL INFLUENCE

Paul F. Lazarsfeld and Herbert Menzel

This paper, like the two preceding it, is concerned with the problem of how communication has an effect, especially the kind of effect that is exerted through personal communication as compared with that which is exerted through the mass media. Who are the "influentials" and "opinion leaders" who help to set styles in ideas and behavior? How is their influence exerted? These are among the questions discussed below.

Professor Lazarsfeld, one of the founders of communication research, was born in Vienna in 1901. He was educated in Austrian schools and earned his Ph.D. at the University of Vienna. In 1933, he came to the United States on a Rockefeller Foundation traveling fellowship, and has made his great career in this country—at Princeton, where he was director of an office of radio research from 1937 to 1940, and then at Columbia, where he founded the Bureau of Applied Social Research and served as its first director. For many years he was professor of sociology at Columbia, and has several times been department chairman. Since 1962 he has been Quetelet Professor of Social Science at Columbia. He is the author and editor of a long list of distinguished and influential books, among them: Radio and the Printed Page *(1940)*; Radio Research, 1941 *(1941)*, Radio Research, 1942–43 *(1944)*, Communications Research, 1948–49 *(1949) (the last three edited jointly with Frank Stanton)*; The People's Choice *(with Berelson and Gaudet, 1944)*; Radio Listening in America *(with Patricia Kendall, 1948)*; Voting *(with Berelson and McPhee, 1954)*; Mathematical Thinking in the Social Sciences *(1954)*; Language of Social Research *(with Rosenberg, 1955)*; *and* Personal Influence *(with Katz, 1956)*.

Dr. Menzel earned a Ph.D. in sociology from the University of Wisconsin, and is on the staff of the Bureau of Applied Social Research at Columbia. His best-known publications are in the field of diffusion and adoption. With James S. Coleman and Elihu Katz, he is the author of a book on the processes of diffusion and adoption among physicians.

Publication No. A-366 of the Bureau of Applied Social Research, Columbia University.

The ability of the mass media of communication to reach large audiences and to make an impact on these audiences became subjects of systematic research in the 1920's. This research reached the proportions of an industry by the end of the 1930's under the impetus of three events: the addition of the film and especially the radio to the previously existing printed media of mass communication; the great expansion and systematization of advertising in the United States; and the apparent success of totalitarian dictatorships in the use of mass media propaganda to bring about apparent dramatic changes in the attitudes of their own publics and sometimes of foreign populations. In all three respects, the power of the media of communication, whether seen as a threat or an opportunity, was regarded as residing precisely in their *mass* character—in the ability of an editor at a central desk or a speaker at a single radio station to reach thousands and millions of readers or listeners simultaneously and to affect their decisions and attitudes.

Subsequent research has shown this image of the simple, direct effect of the mass media to be a great oversimplification. Paradoxical as it may seem, the closer one observes the workings of the mass media, the more it turns out that their effects depend on a complex network of specialized personal and social influences. In the twenty-odd years since the 1930's, mass communications research has experienced a rich development in many directions and a great refinement both in understanding and in techniques of research. Many of the problems received their first clarification in studies undertaken at Columbia University. We shall discuss these studies first, because the story shows how an idea develops from one inquiry to the next. Then we shall discuss the ways in which other scholars have broadened and enriched the original set of ideas.*

* With regard to much of the Columbia material, we draw heavily on an earlier review, where additional details may be found: Katz, 1957.

Opinion Leaders and Their Characteristics

The turning point in this research tradition can probably be identified with a study of voters in the 1940 presidential elections (Lazarsfeld *et al.*, 1948). On the one hand, this study confirmed and supplemented certain regularities of mass media exposure that had by then become quite firmly established. Thus, most of those who learn of an event through one channel, for example, a given magazine, are also likely to learn of it through a second and third channel, for example, a second magazine and a radio broadcast. More educated people attend to more of the mass media. People overwhelmingly select for their attention statements of those opinions with which they already agree. News and opinions about an issue are paid most attention to by those who are most interested in the issue—and that usually means those whose minds are already made up. A paradoxical result of this is the fact that those who read most and hear most about an issue are the ones whose opinions and intentions are least likely to change.

But while the 1940 study confirmed these familiar principles, it also broke new ground. It was the first to take advantage of the so-called panel method, whereby the same people are interviewed repeatedly in the course of an election campaign so as to make possible the detection of changes in their attitudes. This study went to great lengths to determine how the mass media brought about such changes. To our surprise we found the effect to be rather small. We gained the impression that people appeared to be much more influenced in their political decisions by face-to-face contact with other people—members of their family, friends and neighbors, and people with whom they worked—than by the mass media directly.

Detailed case studies helped us to understand the peculiar advantages face-to-face communications have over and against the mass media. They are somewhat more likely than the mass media to reach people who are as yet undecided. Personal influence about an issue is often exerted unexpectedly as a sideline or marginal topic in a casual conversation. It therefore is more likely to "get through" to the undecided or the opposed; mass media messages, by contrast, are more often approached with an awareness of their purposefulness. Face-to-face contact is also more flexible and provides immediate response to instantaneous feedback. A person, unlike a mass medium, is likely to raise issues and arguments of immediate personal relevance to the listener. And finally, when someone yields to personal influence in making a decision, the reward in terms of approval is immediate and personal.

We therefore paid special attention to those who claimed to have been asked for political advice or to have tried to convince someone else of their own political ideas. Contrary to previous belief, these "opinion leaders" were found not to be particularly concentrated in the more educated classes or the more prestigeful positions in the community, but were almost evenly distributed through every class and occupation. They were, however, more interested in the election than the average citizen was and were considerably more exposed to the radio, the newspapers, and magazines. It was, therefore, suggested that communications flow in two steps: from radio and print to opinion leaders, and from these to the less active section of the population.

The importance of personal influence had emerged as a surprise in the analysis of the 1940 voting study. Later studies were able to make better advance preparations to investigate this aspect. One of these studies asked a sample of residents in a small town in the state of New Jersey to name the people to whom they turned for advice regarding a large variety of matters (Merton, 1949). Those who were designated four times or more were considered opinion leaders and were subsequently interviewed. This procedure led to the important conclusion that opinion leadership is not a general characteristic of a person, but is always limited to particular issues. Those who were opinion leaders with regard to many local affairs of the community were not the same as those who were leaders of opinions regarding national political issues. It was these latter, the "cosmopolitan opinion leaders," who differentiated themselves by being readers of national magazines and listeners to broadcasts dealing with public affairs. Opinion leadership thus appeared to be specific to a given field.

To test these ideas in a subsequent study, we deliberately sought out people who had recently changed their opinions or who had otherwise made new decisions in four very different fields: political opinions, the purchase of groceries for the household, fashions, and motion picture preferences (Katz and Lazarsfeld, 1955). Indeed, there was very little overlap of leadership: a leader in one sphere was not especially likely to be influential also in another, unrelated, sphere. Once again it turned out that personal influence played a more frequent and more effective role than any of the mass media. And once again, the influential people proved to be fairly evenly distributed through all educational and income classes, and in general not to be very different from those whom they had influenced. Only with regard to political opinions was there even a moderate degree of concentration of leadership in the highest socioeconomic status.

There thus seemed to be no single trait that would predispose a person to opinion leadership in all fields. Nevertheless, the traits which characterized the opinion leaders in each of the four fields had certain things in common. They placed the opinion leaders in all four fields in analogous positions vis-à-vis their respective fields. There was, first of all, interest: opinion leadership was once again found most frequent among those with the strongest interest in the subject matter concerned. But even with interest held constant, opinion leaders were marked by certain social characteristics: (1) they occupied positions which their communities regarded as giving them special competence in the matter at hand: thus, older women with large families were looked to as marketing advisors by virtue of their greater experience; (2) they were accessible and gregarious and knew many people, especially among those interested in each of the four subject categories;

(3) they had contact with relevant information coming from outside their immediate circle. This last was perhaps the most important characteristic, and it was exemplified by outside acquaintances, by frequent visits to other cities, or by disproportionate attention paid to mass media such as magazines, newspapers, and broadcasts. A new discovery was that opinion leaders were particularly likely to be exposed to the media appropriate to the sphere in which they led. For example, fashion leaders read more fashion magazines, leaders of opinion regarding motion pictures read more movie magazines, and so on.

In the presidential campaign of 1948, an additional study of voting added further details (Berelson *et al.,* 1954). Opinion leaders were found, more often than others, to be consistent supporters of their respective party's position on all possible issues. They also proved to belong to more organizations, to be more likely to know workers for the political party, and to be more frequently exposed to the mass media of communication. In spite of this fact, however, they were also more likely than non-leaders to have sought information and advice from yet other persons. This points to an important modification in the original hypothesis: we deal not simply with a two-step flow—from the mass media, through opinion leaders, to the general public—but rather with a multi-step flow of communications—from the mass media, through several relays of opinion leaders who communicate to one another, to the ultimate followers.

In one respect, the personal influence of opinion leaders was shown to be just like that of the mass media themselves: people tend to have discussions primarily with others who agree with them. Most political discussion, it turns out, goes on among people of like characteristics, of similar age, similar occupation, and similar political preference. When respondents were asked at the end of the campaign what had been talked about in their most recent political discussions, it was revealed that only a small proportion of the discussions involved any degree of argument between the discussants.

This helps to explain why the people who have most political discussions are the least likely to change, in spite of the fact that most of the few changes that do occur are attributable to face-to-face discussions about politics. An equally important reason for this apparent paradox is that political discussions, like political mass media exposure, are most frequent among those who are most interested in politics, and these usually are the ones whose minds are most firmly made up to begin with.

The changes that did occur were found to be most frequent among those who had had discussions with persons favoring the party opposed to the one for which they had originally intended to vote. Changes were also more frequent, the greater the opposition to the individual's original voting intention among family members, friends, or fellow workers.

Communications in Specialized Groups

Once the general idea of personal influence was established, our research focused increasingly on communications reaching and taking place within very specialized groups of people. Our suppositions regarding the characteristics of opinion leaders and of their importance as mediators of mass media influence found confirmation in quite diverse fields.

For example, new kinds of opinion leaders seem to come to the fore as traditional folk communities make the transition to the modern industrial world. When this happens, the community's traditional elders tend to lose much of their pre-eminence to individuals whose position allows them to act as pipelines to the great world outside—provided that they have not become alienated from the local community. Studying village life in Lebanon about ten years ago, for example, we found that the traditional leaders were experiencing an inadequacy which limited the range of their influence and, for the first time, made leadership dependent on the role of the young (McPhee and Meyersohn, 1951; see also Lerner and Pevsner, 1958, pp. 185–196). Those who became opinion leaders were chiefly young people who met our three criteria of (1) holding positions of competence, (2) being accessible and gregarious, and (3) having contact with outside sources of information. They were young men who had obtained some education (position of competence). However, they only achieved opinion leadership status provided they also remained in contact with their families and illiterate neighbors (remained accessible and gregarious). A further distinguishing characteristic of this new kind of influential was his mobility. The "trip to the city" made the difference (contact with outside information).

Perhaps this was the beginning of a specialization, eventually not unlike the one in the small New Jersey town which was mentioned earlier: for the old leaders in the Lebanese villages still controlled the local and the traditional areas of opinion and action, while the new "cosmopolitans" went about relating the village to the outside world.

Physicians, for another example, were found to depend largely on the word of respected local colleagues in deciding whether to accept suggestions made in the professional journals and in advertising media (Coleman *et al.*, 1957, 1963). The local colleagues whom they most frequently sought out as advisors were specialists (occupied positions of competence) who read much and maintained a wide range of contacts with the medical profession in other cities (contact with outside information) and yet identified with and participated in the local medical community (were accessible and gregarious).

Let us now describe this study of physicians in some detail, because it represents a new combination of research methods. The previous studies had investigated only those interpersonal relationships that were mentioned as having played a role in a particular decision. This time, the investigators first determined what interpersonal relations existed among the physicians, and only then asked which of these relations had played a role in the physicians' decision to adopt a new drug. Doctors were, for example, classified according to the number of colleagues who said they habitually discussed cases with them, or who said that they often saw them socially. This made it possible to investigate whether or not the more integrated doctors—those who had professional or social contact with many of their colleagues—adopted the drug earlier than others. (Pharmacists' records of the prescriptions written by each physician were examined for this purpose.) It was also possible to determine whether or not those doctors who had contact with one another had similar drug use patterns. The answer turned out to be yes to both questions, but with many interesting qualifications. For one thing, the better-integrated physicians not only adopted a new drug earlier than others, but they also did so at a rate which increased from month to month. It appeared that those doctors who had been "converted" to the new drug during its first month on the market passed this conversion on to some colleagues during the second month. These new recruits, in turn, went out and converted additional doctors during the third month, and so on. But among the less well-integrated physicians, use of the new drug spread from month to month at a nearly constant rate.

It was possible to measure the power of this conversion or contagion process in greater detail by determining the extent to which those physicians who did talk to each other reached their decisions to use the new drug simultaneously. It turned out that the use of the new medicine spread through the community of doctors in several distinct phases, with personal influence playing a different role in each phase. At first the new practice was passed on primarily between doctors who regarded each other as professional conversation partners. Then it was passed on primarily between doctors who were friends and were intensely involved in these friendship relations. In the third phase, the influence at last reached doctors who were relatively isolated. And in the final phase, individual adoptions occurred without reference to the interpersonal relationships of the doctors concerned.

To be sure, the more formal media of communication, such as professional journals and professional meetings, also played an important role throughout. This was shown objectively by the fact that those physicians who, for example, read a great many medical journals tended to adopt the new drug earlier than their colleagues. Nevertheless, contact with colleagues appeared to be the most influential factor. No doubt commercial media and representatives also played an important role; they were credited with much influence by the physicians themselves. However, they apparently reached all physicians in such abundance that exposure to them does not differentiate between early and late users (or non-users) of the new drug in question.

We may want to pause for a moment to ask why it is that conversations with local colleagues should play such a large role here. To be sure, we have listed earlier some advantages which face-to-face communication has over the printed (or broadcast) word in matters of political persuasion: persons of opposing views are more easily reached by face-to-face communication; they are less defensive when reached; they experience the immediate gratification of the interlocutor's approval if they come to agree with him; and so on. But now we deal with what would appear to be a matter of cold, objective, scientific information, where expert judgments and specialized sources of information are abundantly available. The explanation of the importance of the word-of-mouth in this context must be sought in the often underestimated uncertainty of medical knowledge and, paradoxically, in the very abundance of formal information just mentioned. The formal channels are choked by such an avalanche of professional journals, meetings, conferences, and so on, not to mention commercial messages, that the physician cannot possibly cope with them, even to the extent of selecting wisely what to pay attention to. In this baffling situation, it is not surprising that he turns to his local colleagues to share the burden of screening and evaluating.

These considerations also apply, with some modification, to another group of specialists to which communication research has recently turned its attention: research scientists

(Menzel, 1958, 1959, 1960). These face an overwhelming task of keeping up with an output of new research data that increases in rate from year to year. There exists a great multitude of official media of communication in each of the sciences—scientific journals, books, meetings, conferences, and so on. Scientists spend a great deal of their time attending to these officially established channels of communication. Nevertheless, interviews with scientists have revealed that a good deal of the information which proves to be of significance to them comes to their attention in unplanned and unexpected ways. It reaches them during activities undertaken, and on occasions sought out, for quite different purposes. The scientist may, for example, search the literature for one item of information and stumble across another which proves useful; he may see a fellow scientist to borrow some equipment and be rewarded by an idea which enables him to solve a problem; he may serve on an award committee and unexpectedly gain a new perspective on the trends in his field.

Why should this informal manner of learning of new developments be so prevalent? Part of the reason must be sought in the nature of specialization, especially in the higher reaches of "pure science." The basic researchers not only specialize to a high degree, but there is no simple, uniform set of categories by which their specialities are defined; each uses a different way of slicing the pie. The journals and other official channels of communication also specialize, but their specializations cannot possibly correspond to all of the multiple ways by which scientists define their fields. As a result, much of what is important to any given scientist is dispersed over journals or other media that are ostensibly devoted to specialties quite different from his own. The researcher therefore must depend largely on friends who work in adjoining specialties, and yet know what is of interest to him, to flag pertinent materials for him.

If this is true, it becomes imperative to consider the information network as a system and not merely as an aggregate of information-dispensing or information-consuming individuals. What is little better than an accident from the point of view of the individual may well emerge as an expected occurrence from the larger point of view. For while there is only a small likelihood that any accidentally obtained piece of information will be of use to the individual scientist who obtains it, the likelihood that it will be of interest to at least one of his colleagues in the department is much larger. If enough members of a given research group are plugged into branches of the professional grapevine through consultantships, committee service, personal visits, and so on, they may collectively be able to assure themselves of access to a good share of the news about work in progress that interests them individually. The formal and organized means of communication serve the scientist most efficiently when he knows precisely what he is looking for, when he needs the answer to a specific question. But when it comes to bringing scientists into contact with information whose significance for their own work is not anticipated, then the informal and accidental means of communication are indispensable.

Multiple Functions and Multiple Channels

Another important idea that gradually emerged in the course of the last twenty years of mass communication research is that effective communication demands the performance of several different functions, which are often best served by different channels. Originally, the problem of the multiplicity of channels was recognized merely in the question: Which of the several available channels is most important and most effective? In a more crude sense, the questions were: Shall we put our advertisement into the newspaper or on the radio? How many additional people can we reach by putting them both into the newspapers and on the radio?

When this problem was approached by asking people which of the several channels had influenced a given decision of theirs, it proved impossible to divide people into those who had been influenced, for example, by newspapers and those who had been influenced by radio; for most people had been influenced by several channels. It was suspected that perhaps some channels typically called the individual's attention to the availability of a choice, and that others convinced him that this choice was all right, safe, or legitimate, while still other channels "triggered off the action" by giving him specific instructions on how to execute his decision.

This idea received further confirmation when more careful interviewing in several studies showed that the various channels not only affected individuals in certain typical combinations but also seemed to impinge upon them in certain typical sequences over time. Thus, for example, in the study of new-drug adoption by physicians referred to previously, it was found that the physician's first contact with information about a new drug was typically through a commercial source —in the form of a salesman or advertisement—while the additional information he received later was increasingly likely to be from professional sources (Coleman *et al.,* 1963, Chapter VI).

The idea that different channels of communication play qualitatively different roles in leading persons to a decision finds very strong support in sociological studies of American farming communities. With this we turn from our review of the way we at Columbia University delved into "the part played by people in the flow of mass communications" to an even briefer mention of some of the important contributions that have been made to this line of investigation elsewhere.*

Rural sociologists had noted that different channels of information were associated with the various phases of a decision process. The time during which a farmer gradually made up his mind to use a new, recommended fertilizer or a new kind of seed was divided into phases: in a first phase, the farmer became aware of the innovation; in a second phase, his interest in it became aroused; in a third phase, he reached

---

* Much of the older work that is relevant to this field, although only seldom devoted directly to it, is extensively reviewed in Katz and Lazarsfeld, 1955.

the decision to accept the new practice; and finally, he actually tried it out (Beal *et al.*, 1957). Retrospective interviewing showed that certain kinds of channels of information were typically associated with each of these phases. For example, farm magazines most frequently played a role during the awareness phase, neighbors during the two subsequent phases, and printed pamphlets explained the details of the application of the new practice during the trial phase (Wilkening, 1956; Rural Sociological Society, 1955). The most conclusive of all these studies interviewed not only farmers who had adopted the practice in question but also some who had finally rejected it after going a certain distance along the process of considering it. It was found not only that different channels had once again played the most important role in different phases, but, in addition, that those farmers who had, so to speak, been exposed to the "wrong" channel at a given phase —for example, who had been made aware of the new practice by their neighbors rather than by the farm magazines—had been more likely eventually to reject the innovation than those who had been reached by the "right" channel at the "right" phase (Copp *et al.*, 1958).

Climates of Opinion

But the studies of American farmers also furnish parallels to other aspects of the work reviewed above and add new dimensions to it. We had found that voters, housewives, doctors, and scientists depend heavily on personal advice for the development of new ideas. Other investigators had noticed even earlier that the same is true for farmers when they have to decide whether to adopt new farming methods. In the previous paragraph we discussed the place of such personal advice in the sequence of influences to which farmers are exposed. But who are the persons who exert this influence? The opinion leaders in this context once again revealed the characteristics of gregariousness at home and contact with information coming from the world outside —in the form of reading of farm journals, and of frequent trips outside of the local community. (For a summary, see Rural Sociological Society, 1952.) In addition, the opinion leaders in some farming communities had been found to be individuals particularly hospitable to innovations, but studies in other communities denied this. It was finally realized that farming communities, like other communities, differ in the value they place on efficiency and science and, hence, on innovations. Farming communities were therefore classified for research purposes according to their tendency to adopt new farming practices. It turned out that in areas of high adoption, those farmers from whom other farmers obtained farming information had higher adoption rates than farmers in general; but in areas of low adoption, the adoption rates of the leaders were similar to adoption rates for farmers in general (Marsh and Coleman, 1956). Both facts thus are examples of the general maxim that leaders do not deviate very far from the norms of the groups they lead; if anything, they live up to them with special consistency. As one of our colleagues has stated it, "Those men can better lead who are traveling the same road as their followers but are a little ahead."

More generally speaking, these results point up the importance of the climate of opinion which prevails in the groups with which members of a target audience are affiliated. Studies have also shown that the mere fact of whether a recipient is affiliated with groups at all affects the receipt of messages, their impact, and the way they will be used by the receiver. For example, it was found in one study that children who were well integrated in groups of their age peers listened to quite different radio programs than those who were more used to playing by themselves. Moreover, when children in these two categories did listen to the same radio programs, they made quite different uses of them: the integrated children tended to pick up from adventure programs ideas for games to play with their friends, whereas the isolated children found in them material for daydreams (Riley and Riley, 1955).

More serious in its application, but related theoretically, is a finding made toward the end of World War II: Allied propaganda appeals to Nazi soldiers frequently became effective only after the military situation had led to the dissolution of primary groups in the German army to which the soldiers had belonged (Shils and Janowitz, 1952). In a similar vein, it was found during the war in Korea that North Koreans would accept and act upon United Nations propaganda once they found themselves cut off from the groups to which they had belonged, although the same propaganda had had no effect upon them while they were still members of cohesive units (Klapper, 1960).

More on the Role of Opinion Leaders

Several researchers have gone beyond the Columbia studies in singling out conditions under which opinion leadership is important. We referred above to the situation in folk communities in the process of transition to the modern urban-industrial world. We described how the established leadership by community elders is partially supplanted by younger people who perform the new function of relaying outside news to the community, and show all the other general earmarks of the "opinion leader" as identified in this country. A similar situation was found to prevail among those immigrants who had come to Israel in more or less compact groups from pre-industrial settings in certain Middle Eastern countries (Eisenstadt, 1952; 1955, pp. 190–196). These groups arrived in Israel with their traditional leaders, whose teachings they assembled to hear in synagogues, at arbitration courts, and at assemblies in traditional schools. Sooner or later the position of these leaders is encroached upon by an increase in informal transmission, by different types of opinion leaders for different topics, and by much wider participation of the rank and file in the discussion of important matters. This, according to the author, happens at the point when the traditional leaders can no longer effectively relate their groups to the larger social system, so as to aid members to achieve status in the new environment, mediate values of the new country, and communicate an understanding of the nation's institutions.

During the 1950 congressional election campaign, an attempt was made to characterize the persons most susceptible to the influence of formal and informal leaders (Lowe and McCormick, 1956). The respondents were asked to select three "formal leaders" whose opinions they valued most from a list of sixteen persons prominent in the state where the study was carried out. As informal leaders they were asked to name persons they knew and associated with, and to whom they would go with questions about the election. Ninety-two per cent of the replies referred to family members, friends, neighbors, and co-workers, who tended to be male, slightly older, slightly better educated, and slightly higher placed occupationally than the persons who named them, and to belong to markedly more organizations.

The attitudes people imputed to these leaders were compared with their own attitudes and with changes in these attitudes that occurred in the course of the campaign. Early in the campaign, persons were asked for their opinions on three election issues, and for the corresponding opinions that they believed were held by their formal and informal leaders; after the election, they were asked for their own opinions once more. Most imputed to their leaders opinions like their own; those who saw a discrepancy between their own opinions and that of some of their named leaders tended in the course of the campaign to change their opinions in the direction of those imputed to their leaders. This effect was about equally strong for formal and informal leaders and was strongest among the politically most apathetic. However, adjustments to the perceived opinions of informal leaders occurred most strongly among the least-educated and oldest segments; to those of formal leaders, among the moderately educated and youngest groups.

Of the many other election campaign studies that have been carried out, we will mention here only one of the most recent ones. During the 1960 presidential campaign, the candidates held a series of debates on television. A study of voters' attitudes before and after the first of these debates confirms a number of the earlier generalizations (Deutschmann, 1962). Those who had already made a party choice were more likely to get exposed to the debate or to talk about it. Of those who did get into conversations about the debate, only 11 per cent talked with persons of views contrary to their own, while 47 per cent talked to persons of the same views as themselves, and 42 per cent talked in groups of mixed opinions. The reader will recognize the parallel to our findings (cf. p. 99 above) (1) that people tend to have discussions primarily with others who agree with them, and (2) that political discussions are most frequent among those whose minds are most firmly made up to begin with—because frequent discussions as well as firm convictions are corollaries of a strong interest in politics. By these two facts we had accounted for our otherwise surprising finding that the people who have the most political discussions are the least likely to change their attitudes. The 1960 study similarly found that there was less change in vote intention among those who had talked about the TV debate (19 per cent of them changed) than among those who had not (30 per cent changed).*

When it comes to the mere conveying of information, rather than the exerting of influence, the special features of face-to-face contact would not seem to play such an important role, and one would expect person-to-person communication to be relatively less important. The direct impact of the mass media should be correspondingly greater. The aforemen-

---

* There was more change in vote intention among those who had been exposed to the debate through the mass media than among those who had not; however, the sample of the latter was very small (twenty-three).

tioned TV debate in 1960 came to the attention of 34 per cent of an urban sample through mass media only, to 4 per cent through word-of-mouth only, and to 52 per cent through both. (Ten per cent had not heard of it at all). News of the sudden death of a prominent United States senator in 1953 reached most members of another urban sample through the mass media, with radio being by far the most important channel (Larsen and Hill, 1954). Interpersonal communication was more powerful in the subsample from a university community, where it accounted for 35 per cent, than in a laboring community (17 per cent).

Experiments in Mass Communication

In the more recent past, systematic experiments have been developed to clarify further the role of personal communications as they tie in with more formal media. In one extended field experiment, leaflets were dropped from airplanes containing brief announcements; then people were interviewed as to whether, when, and how they had learned of them (De-Fleur and Larsen, 1958; Dodd *et al.*, forthcoming).

The investigators sought to determine the precise extent to which the spread of simple messages in a community would be determined by such factors as population density, time of day, number of individuals who were given the message at the start, and so on. The greater the number of leaflets per capita which had been dropped, the larger the number of people who had received the message when interviewed three days later. The amount of the increase was found to have followed a simple mathematical equation. The fewer the leaflets dropped, the more important had been the role of interpersonal diffusion in the whole process, with children playing an unexpectedly important role in passing the message on.

Another set of experiments inverted the whole problem of personal influence. The effect of the mass media on certain people was found to depend upon the kind of other people to whom they intended to tell what they were hearing or reading. The anticipation of an audience thus alters the way in which information is gathered and reformulated. The topics remembered, for example, from a speech on teachers' salaries were affected by the kind of audience to which the listeners were supposed to report subsequent to their own exposure: when told that they would be expected to summarize the speech for a group of teachers, they remembered more of the arguments in favor of raising salaries; when told they would be expected to report to a citizens' group dedicated to economies, they remembered more of the arguments against raising salaries (Bauer, 1958).

Thus, personal relations and mass media interact in many ways, sometimes reinforcing, sometimes modifying each other. Much has yet to be learned about these complicated processes. But there can be little doubt about their practical importance and their social significance.

SUGGESTIONS FOR FURTHER READING

1 BAUER, RAYMOND. "The Communicator and the Audience," *Journal of Conflict Resolution*, 1958, 2, pp. 67–77.

2 BEAL, GEORGE M., ROGERS, EVERETT M., and BOHLEN, JOE M. "Validity of the Concept of Stages in the Adoption Process," *Rural Sociology*, 1957, 22, pp. 166–168.

3 BERELSON, BERNARD, LAZARSFELD, PAUL F. and McPHEE, WILLIAM N. *Voting: A Study of Opinion Formation in a Presidential Campaign.* Chicago: University of Chicago Press, 1954.

4 COLEMAN, JAMES S., KATZ, ELIHU, and MENZEL, HERBERT. "The Diffusion of an Innovation Among Physicians," *Sociometry*, 1957, 20, pp. 253–270.

5 ———. *Doctors and New Drugs*, in press (1963).

6 COPP, JAMES H., SILL, MAURICE L., and BROWN, EMORY J. "The Function of Information Sources in the Farm Practice Adoption Process," *Rural Sociology*, 1958, *23*, pp. 146–157.

7 DeFLEUR, MELVIN L., and LARSEN, OTTO N. *The Flow of Information.* New York: Harper & Brothers, 1958.

8 DEUTSCHMANN, PAUL J. "Viewing, Conversation, and Voting Intention," in Kraus, Sidney, ed., *The Great Debates.* Bloomington, Ind.: Indiana University Press, 1962.

9 DODD, STUART C., RAINBOTH, EDITH, and NEHNEVAJSA, JIRI. *Revere Studies in Interaction.* Forthcoming.

10 EISENSTADT, S. N. "Communication Processes among Immigrants in Israel," *Public Opinion Quarterly*, 1952, *16*, pp. 42–58.

11 ———. *The Absorption of Immigrants.* Glencoe, Ill.: The Free Press, 1955.

12 KATZ, ELIHU. "The Two Step Flow of Communication: An Up-to-Date Report on an Hypothesis," *Public Opinion Quarterly*, 1957, *21*, pp. 61–78.

13 KATZ, ELIHU, and LAZARSFELD, PAUL F. *Personal Influence: The Part Played by People in the Flow of Mass Communications.* Glencoe, Ill.: The Free Press, 1955.

14 KLAPPER, JOSEPH T. *The Effects of Mass Communication.* Glencoe, Ill.: The Free Press, 1960.

15 LARSEN, OTTO N., and HILL, RICHARD J. "Mass Media and Interpersonal Communication in the Diffusion of a News Event," *American Sociological Review*, 1954, *19*, pp. 426–433.

16 LAZARSFELD, PAUL F., BERELSON, BERNARD, and GAUDET, HAZEL. *The People's Choice.* New York: Columbia University Press, 1948.

17 LOWE, FRANCIS E., and McCORMICK, THOMAS C. "A Study of the Influence of Formal and Informal Leaders in an Election Campaign," *Public Opinion Quarterly*, 1956, *20*, pp. 651–662.

18 LERNER, DANIEL, and PEVSNER, LUCILLE. *The Passing of Traditional Society.* Glencoe, Ill.: The Free Press, 1958.

19 MARSH, C. PAUL, and COLEMAN, A. LEE. "Group Influences and Agricultural Innovations: Some Tentative Findings and Hypotheses," *American Journal of Sociology*, 1956, *61*, pp. 588–594.

20 MCPHEE, WILLIAM N., and MEYERSOHN, ROLF B. "The Radio Audiences of Lebanon." New York: Bureau of Applied Social Research, Columbia University (mimeographed), 1951.

21 MENZEL, HERBERT. "The Flow of Information among Scientists: Problems, Opportunities, and Research Questions." New York: Bureau of Applied Social Research, Columbia University (mimeographed), 1958.

22 ———. "Planned and Unplanned Scientific Communication," in *Proceedings of the (1958) International Conference on Scientific Information*, pp. 199–243. Washington, D.C.: National Academy of Sciences, 1959.

23 ———. "Review of Studies in the Flow of Information among Scientists." New York: Bureau of Applied Social Research, Columbia University (mimeographed), 1960.

24 MERTON, ROBERT K. "Patterns of Influence," in Paul F. Lazarsfeld and Frank Stanton, eds., *Communications Research, 1948–49*. New York: Harper & Brothers, 1949.

25 RILEY, MATILDA W., and RILEY, JOHN W., JR. "A Sociological Approach to Communications Research," *Public Opinion Quarterly*, 1955, *15*, pp. 445–460.

26 Rural Sociological Society, *Sociological Research on the Diffusion and Adoption of New Farm Practices*. Lexington, Kentucky: Kentucky Agricultural Experiment Station, 1952.

27 Rural Sociological Society, North Central Regional Subcommittee on the Diffusion of New Ideas and Farm Practices, *How Farm People Accept New Ideas*. Ames, Iowa: Iowa Agricultural Extension Service, Iowa State College, Special Report No. 15, 1955.

28 SHILS, EDWARD A., and JANOWITZ, MORRIS. "Cohesion and Disintegration in the Wehrmacht in World War II," *Public Opinion Quarterly*, 1952, *12*, pp. 280–315.

29 WILKENING, EUGENE A. "Roles of Communicating Agents in Technological Change in Agriculture," *Social Forces*, 1956, *34*, pp. 361–367.

9 THE EFFECTS OF TELEVISION ON CHILDREN

Eleanor E. Maccoby

Is television really harmful to children? In particular, has it anything to do with juvenile delinquency? Does the violence on some television programs make children behave more violently? Are there some kinds of children on whom television is more likely to have bad effects? Questions like these are important to us all. Mrs. Maccoby gives careful, reasonable answers to them in the following paper.

Mrs. Maccoby was born Eleanor Emmons in Tacoma, Washington, in 1917. She graduated from the University of Washington and took both M.A. and Ph.D. degrees from the University of Michigan, where she was for a time a study director at the Survey Research Center. During most of the 1950's she was a lecturer in social psychology and a research associate in the Laboratory for Development at Harvard. In 1958 she was appointed to Stanford, where she is an associate professor of psychology. She is the author of a number of articles on children and television. Her best-known book is Patterns of Child Rearing *(with Sears and Levin, 1957), and she is one of the co-editors of a much-used reader in social psychology. She is Mrs. Nathan Maccoby.*

It has been only a little over a decade since a television set became part of the equipment of almost every American home. And television has become available to large numbers of Western European families only during the last few years. In this short time, television has brought great changes in the lives of children.

When television began, there were great hopes and great fears concerning what its effects might be. The optimists believed that it would educate children in the widest possible sense, giving them an opportunity to learn about science and about human life in other lands in such a fascinating way that learning would be a joy instead of a labor. The pessimists worried about whether too much television would damage children's eyesight, whether it would keep them from healthy active play, whether it would corrupt them by showing too much crime and violence, and whether it would interfere with their study and learning of school subjects.

Despite the fact that television has been in widespread use for only a short time, we do have a new generation of children growing up who have been exposed to television throughout the major part of their lives. So perhaps we do have enough experience with TV now to begin to assess its effects. This will not be an easy task, however, for this generation of children has obviously been exposed to many new influences other than television: these children have grown up in a time when society was recovering from the effects of a major war and living in the shadow of another one; when population has grown rapidly, with crowding of houses and schools and rapid expansion of suburban living; when income and living standards were rising. If present-day children are different from those of earlier generations, different for better or worse, it is difficult to know how much of the credit or the blame to lay at the door of television, considering these many other changing conditions to which they have also been exposed.

How, then, can we go about tracing the effects of television per se? How can we distinguish these effects from those of the other important social conditions which have been influencing children at the same time? One approach is to find children who are growing up with television and compare them with another group of children who are growing up in the same generation and in similar circumstances but without television. It has been possible to find a few situations in which a "natural experiment" of this sort can be done. The most outstanding examples of studies of this kind have been the Nuffield studies, carried out by Dr. Hilde Himmelweit and her colleagues in England, the work of Takeo Furu in Japan, and the studies done by Wilbur Schramm and his colleagues in American and Canadian communities. Another method for assessing the effects of television is to study children before and after they have seen particular programs or groups of programs, to see whether there are changes which occur in their knowledge, attitudes, or behavior which are not found among children who did not see the programs. This approach has been used in an extensive series of studies done in Australia by F. E. Emery and R. V. Thomson, and in individual studies by research workers in the United States, France, and Germany. In this brief paper I cannot, of course, describe and discuss all these studies. Let me simply summarize as best I can what I believe the studies as a whole show.

First of all, let us ask this: When children have unlimited access to television, how much time do they spend watching, and what kind of programs do they watch? If one thinks solely in terms of the number of hours spent, there can be no doubt that television is a vastly important element in children's lives. In the United States, children's viewing rises from an average of about two hours a day at age five to about three hours a day at age twelve to fourteen years—the age when viewing is heaviest. In England and Japan, the average was somewhat less; about two hours daily was average for ten-year-olds in England, for example, when the Nuffield studies were done several years ago. But whether the average viewing time be two or three hours daily, it is a substantial amount of time—more time than children spend on any other single form of leisure-time activity. What do they see during these hours?

Overwhelmingly, children choose to view programs which entertain rather than educate. Their favorite programs are cartoons, Westerns, animal and crime dramas, and family situation comedies. A very large amount of children's viewing time is spent on programs intended primarily for adults. In one study, first graders were devoting 40 per cent of their viewing time to what most viewers would call adult programs, and by the time the children were in the sixth grade, four-fifths of the programs they saw were adult programs.

If we want to consider the effects of the programs children see, it is evident that we cannot consider only the content of the so-called children's programs; we must consider the whole spectrum of adult dramatic programs as well. But perhaps the most important thing to remember concerning children's choice of programs is that they, like their elders, use television primarily for entertainment. Even when educational programs are available, few children watch them if there is a choice of an entertainment program to watch instead. Children who are seriously interested in informing themselves on a topic usually use books, magazines, and newspapers for this purpose. I do not mean to say that children do not learn anything from television, but when they do learn, it is usually incidental to the entertainment value of a program. Television has been called "chewing gum for the eyes," and it is an apt phrase.

But, of course, to know how much time children spend on television, and the kinds of programs they usually watch, does not tell us much concerning the *effects* of this viewing. Does it produce any lasting effects? Or is it only a source of momentary pleasure that has little bearing upon the development of a child's thinking and values? I have been discussing how much time children spend on television, and the kinds of programs they choose. Now let me summarize some of what is known about the effects of this viewing.

First of all, it appears that some of the early worries about the possible harm television might do were unfounded. Television has not damaged children's eyesight. And the best evidence appears to show that it has not made much difference in children's school performance. That is, when children who have TV in their homes are compared with children who do not, there is little difference between them in the extent to which they are interested in their classroom work; the two groups of children do equally well in completing their assignments of schoolwork to be done at home, and they perform equally on tests. The single exception to this conclusion comes from Japan, where seventh-grade boys who had TV in their homes spent less time on homework and suffered somewhat on reading tests by comparison with a matched group of boys who did not have access to TV. This effect of TV on homework and on reading ability was not found in younger boys in Japan and has not been found in studies in other countries.

There is some evidence that television is a stimulating experience to very young children, for children who have access to television during their preschool years come to school with larger vocabularies than children who do not have the experiences provided by television. But the vocabulary differential disappears soon, under the impact of school training, so it appears that the early impetus which television gives to language learning does not constitute a long-range advantage.

Surprisingly, TV does not seem to interfere in any substantial way with the reading of books: children seem to do about the same amount of book-reading regardless of whether or not they have television at home. And there are a number of documented instances in which the presentation of a classic story or play on TV has stimulated interest in reading the original, so that libraries report a suddenly increased demand for the book. But television does clearly have an impact on radio listening and the reading of comic books: these forms of entertainment and television seem to be substitutes for one another, so that when a child begins to spend more time on television, he spends less on radio and comics.

I mentioned earlier that many people have been concerned about the amount of violence and crime children see on television and have wondered whether this might serve to increase the aggression displayed in real life by the children who watch it. Certainly it is true that the average child sees many fist fights, gun fights, and crimes in a week of television. And even in the cartoons which children love so well, fighting is one of the major themes that provides the excitement of the stories. One cannot easily dismiss the possibility that these television episodes are teaching aggressive behavior to children. Yet it is difficult to determine whether this is true. It has been found that the more hours a child spends watching

TV, the more likely he is to express aggressive impulses on personality tests designed to measure hostile feelings. But we do not know whether the many fights children see on television have made these frequent viewers feel more aggressive, or whether, on the contrary, they choose to watch a great deal of television because they are already aggressive and find television fare satisfying to their impulses.

We have the same problem of interpretation when we discover that delinquent children enjoy watching crime shows or fight scenes. It is true that there are a number of well-documented cases in which a young delinquent has imitated crimes seen on television or in the movies. But to say that the programs he watched were entirely responsible for his anti-social behavior would be completely unfounded. The large majority of children who saw the same programs did not imitate the crimes portrayed. And we know that we can almost always discover *other* conditions in the background of a young criminal that predisposed him to crime. Poor home training and influential street-corner gangs are undoubtedly much more potent influences on a child's becoming a good or poor citizen than any number of television programs. But can we say that the television programs have made some difference in their own right?

There are recent studies that bear upon this point. One is by Ivar Lövaas, who wished to discover whether cartoons which emphasized fighting would make children feel more aggressive or less so after viewing the cartoon. He showed an aggressive program to one group of children and some non-aggressive material to another group. Immediately after viewing, each child was given a choice between two toys to play with, one of which was an aggressive toy. If the child turned a lever activating the aggressive toy, he could make two dolls hit each other on the head. The other toy which the child could choose had moving doll figures that did not hit one another. The children who had seen the aggressive cartoon tended to play with the aggressive toy immediately afterward, while the children seeing the non-aggressive cartoon preferred the non-aggressive toy. This finding suggests that viewing aggressive programs serves to arouse children's aggressive impulses in some degree.

A similar result was obtained by Albert Bandura *et al.* These researchers showed some children a movie of an adult hitting and kicking a "bobo doll"—a large balloon-like doll with weights in the feet which can be used as a kind of punching bag. A control group of children did not see this film. On a subsequent occasion, when something had happened to make the children feel irritated and frustrated, the children were taken one at a time into a room which contained a bobo doll as well as other toys. The children who had seen the film imitated quite exactly the behavior they had seen in it; they hit and kicked the bobo doll, while the children who had not seen the film did not do this. In fact, the children who had not seen the film displayed considerably less aggression of all kinds in the test situation.

A third study, by Paul Mussen and Eldred Rutherford, showed that children who had just seen an aggressive animal cartoon were more willing to engage in destructive play involving popping balloons than were children who had not seen the cartoons.

We should note that these studies do not demonstrate that seeing aggressive programs will cause children to become directly aggressive toward other people. In each case, the test involved seeing whether the child would become more aggressive toward a *doll*, and this is a different thing than, for example, hitting another child. But these studies do suggest several things to us. In the first place, seeing aggressive episodes on television serves to arouse aggressive feelings in the viewer. In case any of us have thought of aggressive fantasy as a kind of safety valve that harmlessly releases aggressive impulses, our view would receive little support from these studies, for seeing episodes in which fighting occurred did not "discharge" the viewer's aggressive feelings. In fact, as judged by his subsequent behavior, the opposite occurred. Secondly, children do learn while viewing, for when faced with appropriate conditions later, they will carry out the same actions they saw on the screen.

If this is the case, why don't we see children imitating television heroes more often? As Bandura points out, the crux of the problem lies in whether the child has the same instruments at hand. He won't imitate a burglary in which a safe is broken open because he is never in a situation when he is alone with a safe and a set of burglar's tools at his side. But when he sees a television episode, he has in a sense added an item to his repertoire of *potential* behavior. Whether he will ever carry it out will depend in part on whether he ever encounters a similar situation. And, of course, whether a child imitates what he has seen will depend on many other things, too. It should depend on whether the behavior he sees is consistent with the moral standards of behavior he has been taught, and whether the moral teaching has been effective enough to enable him to overcome an additional bit of temptation.

So we see that even though children seldom watch educational programs, they are nevertheless being educated in a certain sense by the entertainment programs they watch. That is, they are learning things from these programs about how people behave in a variety of situations, and under appropriate conditions they will try these actions out themselves. Can it be said that they are learning attitudes or values or beliefs, as well as actions? A number of efforts have been made to assess the degree to which children's attitudes are influenced by programs they have seen. An early set of studies (by Peterson and Thurstone) was done in the 1930's on the effects of movies in Chicago, and it was clearly shown that attitudes toward a racial minority could be influenced by the way in which a member of the minority group was portrayed in a motion picture. More recently, Alberta Siegel showed that children could form a stereotyped impression of what taxi drivers as a group were like through seeing a program in which a taxi driver played a particular role. It is likely, incidentally, that a story portrayal of actions of someone in a more familiar occupation did not have much influence on children's views about the characteristics of people in that occupation. This is suggested by the findings of the earlier Peterson and Thurstone work: that movies had their greatest effect upon the attitudes of children who had had very little previous experience with members of the minority group being portrayed in the movie.

It is difficult to evaluate the *cumulative* effects of television programs on children's beliefs and attitudes. One can show them a *single* program or movie and question them afterward to discover effects of this one experience upon their attitudes. But an attitude acquired from one program might be counteracted by an opposite attitude presented in the next program the child saw. The greatest impact on attitudes ought to occur in those areas where television programs present the same theme repeatedly, with only slight variations. We can all think of such stereotyped themes—the Western hero who shoots only to punish an evildoer; the martyred wife in the women's afternoon program who bears the burden of adversity secretly and alone so as not to disturb the happiness of others; the dog or horse who knows more than his master and saves a man or child from his own folly; the tough, smart detective who turns the crook over to the baffled police. Is there any reason to be concerned about the cumulative effects of repetitive themes such as these? We have really very little evidence one way or the other. Dr. Himmelweit and her associates in England did find that young girls who see many programs about the problems of adult life become worried about what marriage will be like, and see adult life as filled with one crisis after another involving unfaithful husbands, ungrateful children, illness, and so on. Girls who have not seen so many of the daytime serial plays are more optimistic, more likely to think about the joys and rewards of adult living. Each one of us may have his own opinion about which view is more justified by the facts of modern life, but the point is that television appears to be fostering one view rather than the other.

So far we have indicated that television does have some effects on young viewers, by providing them with models for action which they can imitate when the situation seems to warrant it, and by transmitting certain attitudes and feelings about people and events. There is another kind of effect which has been emphasized in some Australian research. This is the arousal of emotional defenses. Children who see an unusually large number of crime programs do not continue indefinitely to become aroused to an intense emotional state with each program; the process would be emotionally too exhausting. They therefore repress their emotional reactions, with the result that they tend to be somewhat depressed and apathetic after viewing such programs, and seem to be less capable of a sympathetic reaction toward people who are in trouble than are children who do not see a great many crime programs.

I have been listing some of the television effects which research has discovered in some children under some conditions. I would like to emphasize that television is not the only, and not even the major, influence upon children's attitudes and values in most spheres of life. When television presents values and models for behavior which are not consistent with the values parents hold for their children, there is no reason to believe that the influence of television will be paramount. Where children have already acquired knowledge or values on a subject, television seems to have little effect. It is in the area of the unfamiliar, where parents have not yet made clear their own point of view and where the child has little real-life experience to use as a guideline, that television will influence beliefs and attitudes and establish stereotypes.

So far I have been talking as though the effects of television were fairly uniform among all children, yet this is by no means the case. Schramm and his colleagues have insisted upon the point that the child is not a passive entity, being acted *upon* by television, but that he is an active agent, selecting from TV material that which fits his interests and needs best. It is certainly true that some children watch far more television than others, and that the motives which inspire them to do so are various. It has been quite commonly found that the children who spend more than the average amount of time in watching TV appear to be using it as a form of escape from tension. Children who are going through a period of difficultiy in making friends, or of conflict with their parents, spend unduly large amounts of time with television. And it is not surprising that the level of a child's intelligence has a great deal to do not only with how much he watches but with which programs he chooses. His choices are also influenced by the education and taste of his parents. Furthermore, among a group of children who have chosen to watch a given program, some are undoubtedly more susceptible than others to its influence; children differ in what element of a program captures their closest attention, and what details they remember. So in a certain sense, the influence of television is unique with each child.

But I believe it is nevertheless meaningful to consider the over-all effects on children taken as a group. I have tried to summarize the evidence concerning these effects. There are not many satisfactory studies. Those that have been done do point fairly consistently to certain effects: there is reason to believe that children's attitudes and beliefs can be shaped by what they see on television and that emotions and impulses are aroused in the child viewer to match those portrayed by screen characters; it is a reasonable conjecture that the child responds in kind to the emotional states depicted on the screen, whether these emotions be anger, sexual feelings, joy, or self-sacrificing altruism. All this is not too surprising. We have long assumed that dramatic art had other functions than simple entertainment—that great plays have a lasting meaning because, among other things, they serve to interpret reality, to give us new ways of seeing and understanding life experiences. Children appear to be using television in just this way—as one of the sources from which they draw material for organizing and interpreting their experiences. They also use it to prepare themselves for their future lives as students, as marriage partners, as members of a professional or occupational group.

All this means simply that television is part of the total environment that we, as a society of adults, provide for children. Although television is a fairly new element in this environment, it is already providing a share of the influence which shapes children's thoughts and actions. I would assume that this means that informed citizens must be concerned about television, take some responsibility for it, and apply standards of evaluation to it, in the same way they do for other important aspects of the environment that affect children, such as schools, or conditions that affect children's health and safety. There are some complex issues of policy here. There is certainly room for difference of opinion concerning where the responsibility should rest for guiding children's viewing and for bringing standards of taste and quality to bear upon the programs that are offered to viewers. But there can no longer be much doubt that television does constitute an important source of influence on children and as such is a legitimate object of public concern and public action.

SUGGESTIONS FOR FURTHER READING

1 HIMMELWEIT, HILDE, OPPENHEIM, A. N., and VINCE, PAMELA. *Television and the Child*. Published for the Nuffield Foundation. London and New York: Oxford University Press, 1958.

2 SCHRAMM, WILBUR. *The Effects of Television on Children: An Annotated Bibliography of Research* (in press).

3 SCHRAMM, WILBUR, LYLE, JACK, and PARKER, EDWIN B. *Television in the Lives of Our Children*. Stanford: Stanford University Press, 1961.

10 THE EFFECT OF COMMUNICA-
TION ON VOTING BEHAVIOR

Ithiel de Sola Pool

What effect do the mass media have on the results of elections? Does it make any difference in election results if a great majority of the newspapers represent one party? On what kind of information does a voter decide how to cast his ballot? Is personal influence more important than the mass media in the voting decision? These are the kinds of questions introduced by the following paper.

Dr. Pool was born in New York City in 1917. He earned three degrees, including a Ph.D. in political science, from the University of Chicago. He taught at Hobart and William Smith Colleges, then at Stanford, where he was director of a communication research project at the Hoover Institution. Since 1953 he has been at M.I.T., where he is director of an international communication study and a professor of political science. He is co-author of American Business and Public Policy: The Politics of Foreign Trade *and of* The People Look at Educational Television *(in press), author of several monographs, including "The Prestige Papers" and "Symbols of Democracy," and editor of a volume on* Trends in Content Analysis. *He has been a consultant on political communication to a number of governmental and private organizations.*

To the student of human behavior there are many fascinating things about elections, quite apart from who may win. Elections are, among other things, a laboratory. One knows in advance the exact day when an election will occur, which is not true of most human events. This gives the researcher the opportunity to schedule his study, something he cannot do with many other important forms of human behavior. Crises, revolutions, rumors, technological changes, or religious revivals do not announce themselves in advance. Elections do. And one knows not only the date of an election but also the form the outcome will take—a certain number of votes will be cast for A and a certain number of votes for B. So the social researcher has a chance to test his theories and to prove out his methods by making a prediction of what those numbers will be. There need be little doubt afterward whether he was right or wrong. The results of an election are exact and they are measured. They thus provide an opportunity for testing a theory almost as if by a controlled experiment.

It is no wonder that in recent years elections have been a favorite subject of study by social scientists all over the world. In the United States they are studied mainly by public opinion polls, but other techniques are also available. In France, a research technique called "political geography" has been favored. The distribution of the vote by party is plotted on a map along with other characteristics. These characteristics could be crop or economic areas, newspaper circulation figures, or any other items which are found to correlate with party preference. As we said, however, in the United States public opinion polling has become the favored technique of election research. It is being widely used in many other countries, too.

Among the outstanding studies in the United States, one must mention the panel studies done at Columbia University by Paul Lazarsfeld, Bernard Berelson, and William McPhee. A panel study is one in which the same individuals are interviewed every few weeks. In that way the researcher can follow exactly how each individual in the panel makes up his mind during the course of the campaign—how often he has changed it or wavered; what newspapers, magazines, radio or television programs influenced him; and what personal friends influenced him.

The trouble with a panel, however, is that frequent repetition of interviews is expensive, so that most studies which attempt to be nationwide in scope in a country as large as the United States have to cut down on the number of re-interviews. An example of such national election research studies is provided by the work of Dr. Angus Campbell at the Survey Research Center of the University of Michigan.

Similar interesting election studies have been made in England by Professor Robert MacKenzie of the University of London. Other excellent studies have been done in Norway and Sweden. And polling, the basis of such studies, is by now almost as widespread as voting. There are polling organizations now in most countries of the world.

Our subject, however, is not polling or even elections. It is communications and elections.

Elections are particularly interesting research opportunities for the student of communications. They offer him a chance to observe in action the processes of persuasion, of attitude change, of the lining up of sides, of the response of the public to news.

Take, for example, a remarkable fact about elections observable in many countries having a two-party system. The United States is one of these. Year after year, in election after election, the results are very close to an equal division. Neither party ever gets a large majority. In the United States sometimes the Republicans win, sometimes the Democrats, but whichever wins, it is with 51 or 52 per cent of the vote— sometimes 53 or 54—but rarely more than 55 per cent. The situation is similar in England. One may well ask how this happens.

It is certainly not true that subsectors or special groups within the general public divide themselves evenly between two parties. In the United States, for example, Southerners tend to be Democrats. So do factory workers and so do working newspapermen. Businessmen tend to be Republicans, as do college-educated people and newspaper publishers. Indeed, in elections for the past three decades, American newspapers have editorially endorsed the Republican over the Democratic candidate for president, generally by a ratio of something like four to one. But that has had very little influence on the popular vote.

How does it happen that with the public consisting of numerous somewhat one-sided blocs, it winds up so neatly balanced as a whole?

One guess is that the explanation lies in the similarity of the two parties in the United States. It is often hard for visitors from abroad to recognize any difference between the Republican and Democratic parties. But that guess is not right. The parties are very different indeed. If one were to talk to what in Europe are called the "party militants" of each party, or what we call the "party workers," one could not help but notice substantial differences in viewpoints and backgrounds. In England, likewise, while Labor and Conservative governments may be similar in many policies, the people at the annual party congresses differ greatly from each other. It is much the same in the United States, as can be noted at the national party conventions held every four years.

How, then, does it happen that two such different parties end up by being so much alike? The answer is that both want to win, so both adjust their own viewpoints and prejudices and ideologies to hit that point which represents the real opinion of the public.

The two parties do not end up so close to a 50–50 division because they are alike. Rather, they end up alike because each is striving to get that magic 51 per cent needed to win. The result of this competition for the good will of the voters is that the common elements in the programs of the two parties are an extraordinarily good representation of public opinion. The area of consensus of the two parties represents quite exactly the mainstream of public thought.

Persons holding minority viewpoints often claim that the similarity of the two parties deprives the voters of any real choice, but the truth is just the other way around. The choice of the voters has already imposed itself upon the parties before election day. In that way, the average voter in the middle, not the party militant, has the last word.

Now if that is so, then the most interesting question for research is how that voter in the middle makes up his mind.

He is an uncommitted voter, a non-party voter, or a weak party member who is ready to shift. What more can we say about the sort of person he is? What does he read? What causes him to change his vote from election to election?

In the first place, we can say that the number of these people, who hold the balance of power in their hands, amounts to something like a third of the American public—maybe a little more. Another third of the public consists of convinced Republicans, of people who will vote Republican no matter what. The remaining third of the public consists of convinced Democrats, also unlikely to change. The uncommitted third contains many who lean toward the Democrats and a few who lean toward the Republicans, but the people in this group are ready to go either way if they vote at all.

Sometimes, indeed, these fence-straddlers don't even vote because they are less interested in politics than the more committed voters. The independent is seldom a judicious but detached person. Most often he is relatively apathetic, ill-informed, and unpolitical. He is ready to be influenced and must be influenced if he is to take any part in public life, since he is not a self-starter. But what influences him?

In part, this relatively uninvolved voter is influenced by the mass media. In the United States he is reached by newspapers and television in particular. But he is not affected by them in the way one might think. The media do not often convert him. We have already noted, for example, that the media have mostly supported the Republican presidential candidate every four years since 1932, but that the Democrats have won six out of eight times. Why is it that the gigantic and ubiquitous American press appears to have had so little impact? There are several reasons.

In the first place, American media generally make a serious effort to report what both candidates say. The newspapers are not party organs. An increasing number, for example, do what our radio and television do all the time—they give the speeches of each competing candidate exactly equal and parallel space. Most papers don't do that, but they do report both candidates. They never confine themselves to reporting only the views of the man they favor.

Studies of voter behavior have shown that voters look for and read the statements of the candidate they already support. This process of selective perception means that as long as each side has its statement somewhere in the paper, the voter will be reinforced in his views—not converted—even by a paper which supports the other side. The paper may favor one side but if it reports both, the voter can successfully insulate himself against conversion.

A second reason why the media fail to convert many voters is that as people become more interested in politics during the course of an election campaign their views become firmer. Those individuals in the public who are inclined to read about politics or to listen to political speeches are the ones who already have strong views which are not going to be changed in the eight weeks of an election campaign.

On the other hand, the people without strong views who might be converted don't read or listen initially because they are not deeply interested. Of course, as the campaign progresses they become more interested. Thus, ironically, they become both more likely to pay attention and less likely to be converted.

The main effect of a political campaign is to mobilize, not to convert. It revives the lagging interest of the average voter, but as it does so, it also revives whatever latent convictions he already had. In other words, the arguments he reads in the mass media or hears from speakers on television serve on the one hand in part to persuade him, but serve on the other hand to revive his recollection of his lifelong convictions. The latter will often be on the side of the issue diametrically opposite to what is being presented to him.

For these reasons the main effect of the mass media in an election is to draw the election and the issues in it to the public's attention. Once the people have become aroused, their decision in the short run is more highly influenced by personal stimuli than by the mass media. Personal influence by family, friends, and neighbors takes over whenever the media raise a political issue to that level of public attention at which people talk about it face to face. There is a lot of evidence from election studies on the effectiveness of personal influence. For example, election studies show that as a campaign proceeds, families, groups of friends, or other social groupings become more homogeneous. In each of the last three American presidential elections, for example, men and women divided slightly differently, with more women than men voting Republican. But the difference was smaller by the time election day rolled around than it had been in public opinion polls early in the campaign. What happened was that as free public discussion took place, wives converted their husbands or, more often, husbands converted their wives, so that families tended to become more homogeneous.

It was not the newspaper the family was reading which most often determined how they voted. That only got them to talking. It was personal influence in the face-to-face environment which produced the most change. The primary community environment continues to be the place where people's attitudes are formed, even in a society as modern and urban as that of the United States.

In an election campaign in such a society, the mass media are important, too, but not, as we have seen, in the ways they seem to be at first glance. For example, in a free society, the mass media serve as a unifying element. They tie together the many pluralistic face-to-face groups in which attitudes are formed. They bring certain common issues to the attention of all of them. The media serve to place the current issues of politics on the agenda of citizen discussion. They bring citizens who, between elections, are absorbed in private affairs back periodically to the public *agora* to consider affairs of state.

On the other hand, the free mass media—that is, media which allow expression to diverse and opposed views, as the American media do—do not change or control peoples' opinions much in the short run. The influence process in an election campaign is to a greater degree the other way around. Because the opinions of the people are already formed and apparent to the politicians, the latter to some extent repeat in their speeches and platforms, which go out over the mass media, what the public already believes. An election campaign in the United States, while it makes the voters more politically alert, and as a result more partisan, pushes the politicians in just the opposite direction. Their partisan tendencies become moderated as they seek the approval of 51 per cent of the voters.

Thus, in mysterious ways, an election campaign and the mass media coverage of it do somehow serve to give fairly good expression to a basic consensus that exists in the opinions of the people.

The way in which the national consensus expresses itself is most apparent in the most important elections. In the United States these are the presidential elections, which are held every four years. It is these elections which get the most extensive press coverage but in which, for the reasons already noted, the influence of the press is, ironically, at a minimum.

The press in the United States is much more influential in local than in national elections. Americans elect a great many of their officials. In addition to electing a president and vice-president, they elect senators and representatives in the federal government, governors of their states, senators and representatives in the state legislatures, mayors, city councilors, and members of school boards. In addition, in many jurisdictions they elect such other officers as the state attorney general, the state auditor, the lieutenant governor, county commissioners, judges, district attorney, coroner, sheriff, and many others. The result is what is known as the "long ballot." The voter enters the polling booth to mark a ballot covering separate contests for perhaps ten or a dozen different offices and sometimes, in addition, for several referenda. While a national campaign may have focused his intense interest and deep convictions on the top office, he is often without any such internal guidance on the less important offices. He has neither time nor energy to inform himself on all of them. So how he votes on these minor offices is apt to be affected by any information that comes his way about the candidates for them. It is in this situation of low intensity of attention and interest that the endorsement of a candidate by a newspaper is capable of influencing a number of voters.

Note the irony of this situation. The press pays great attention to a presidential campaign. It gives thousands of columns of space to it. Yet the net effect of the press on the voter's choice is small. It gives very little attention to contests for minor local offices. Yet its net influence on the voter's choice for these is substantially greater. In short, the direct influence of the mass media is greatest where they do not stir up public interest to the point of mobilizing the more sluggish but basically more persuasive oral communication system.

This last observation leads to a further point about the influence of the mass media on public opinion. The power of the media to persuade, at least when there is democratic controversy, is very much less than is usually assumed, but their power to inform is enormous. Every morning the daily newspaper brings in a truly unprecedented fare of facts and images from all over the world. Its coverage is very much greater than is possible by any system of oral communication. The daily paper, for example, has no difficulty at all listing the dozen different contests which will appear on the ballot, naming the candidates for each, and giving thumbnail information about the record and policies of each of them. In the modern world, all of us depend upon the mass media for much of the large stock of facts and images which we need to deal with the complexities of life. But what we do with that information is something the media do not control.

In closing, let me mention one particularly dramatic instance of mass communication which did play a very important part in the 1960 presidential election campaign.

There was a series of four television debates between the two candidates, John F. Kennedy and Richard M. Nixon. Until the debates began, a great many voters were undecided. Then came the debates, which brought into the living rooms of most of the American people a face-to-face confrontation of the two aspirants. When they had seen that close-up of the candidates in action, millions of people promptly made up their minds. There were many extraordinary things about the debates. First was their enormous audience. The large majority of all American adults saw at least a portion of this great forum. A second noteworthy thing about the debates was that they caused partisans of both parties to listen to both candidates. The debate broke through that wall of selective attention by which people normally listen only to speeches of the man they already favor. Finally, the debates were noteworthy in that they forced each candidate to meet the issues raised by the other. For all these reasons, the debates became the decisive event of the 1960 campaign, the event which led Mr. Kennedy to win the election.

Debates are not a new thing in American politics. There is an old tradition of debates. Among the famous debates were the Lincoln–Douglas senatorial election debates over a century ago. But never before has it been possible to stage a debate before an entire nation as the audience. Now, with over 90 per cent of American homes having television, major public issues can, on occasion, be presented in town-meeting fashion with the whole country watching.

SUGGESTIONS FOR FURTHER READING

1 BERELSON, BERNARD, LAZARSFELD, PAUL, and McPHEE, WILLIAM. *Voting: A Study of Opinion Formation in a Presidential Campaign*. Chicago: University of Chicago Press, 1954.

2 CAMPBELL, ANGUS, CONVERSE, PHILIP E., MILLER, WARREN E., and STOKES, DONALD E. *The American Voter.* New York: John Wiley and Sons, 1960.

3 LAZARSFELD, PAUL, BERELSON, BERNARD, and GAUDET, HAZEL. *The People's Choice.* New York: Columbia University Press, 1948.

4 POOL, ITHIEL DE SOLA, ed. *Public Opinion Quarterly,* special issue, "Studies in Political Communication," Spring, 1956.

11 TEACHING MACHINES AND PROGRAMED INSTRUCTION

Arthur A. Lumsdaine

Effective education depends on effective communication. In the last few years some of the most interesting communication research has dealt with the processes of teaching and learning, in an effort to improve education through improving its basic communication. The development of programed instruction has been one of the most exciting results of this research. Programed instruction has been called "the first prose ever constructed especially for teaching." In a more extravagant vein, it has been called "potentially the most significant development in instructional materials since the invention of the printed book." The following paper discusses this new method of teaching.

Dr. Lumsdaine was born in Seattle, Washington, in 1913. He took a baccalaureate degree at the University of Washington, and a Ph.D. in psychology at Stanford in 1949. He served as an instructor at Princeton and was an assistant professor at Yale. During the war years he was active in research for the Army, and out of this experience came the book he wrote jointly with Carl Hovland and Fred Sheffield, Experiments on Mass Communication *(1949). He is now a professor of education at UCLA. Before his appointment to UCLA in 1959, he was program director for training and education at the American Institute for Research and, earlier, directed a U.S. Air Force research program dealing with teaching methods and teaching aids. His interest in programed instruction is illustrated by two recent books,* Teaching Machines and Programmed Learning, *edited jointly with Robert Glaser in 1960 for the National Education Association, and* Student Response in Programmed Instruction, *edited in 1961 for the National Research Council, National Academy of Sciences. (Programed is spelled by some practitioners and scholars with one* m, *by others with two* m's.)

During the past five or six years an increasing amount of experimentation has been devoted in the United States to a new way of teaching. This new approach to teaching is called programed instruction. Frequently, though not always, it utilizes special kinds of devices called "teaching machines."

The concept of programed instruction has both a general meaning and a more specific one. In the more general sense, it simply implies a sequence of instruction that is worked out carefully in advance and is then recorded so that it is reproducible. This might include tape recordings, films, or lessons recorded for presentation over television. But by programed instruction, as we use the term today, we mean something more than just a vehicle for presenting knowledge. The "something more" lies in the degree to which we really take responsibility for instructing, rather than merely providing an opportunity to learn.

Organizing a subject in a textbook is a step in this direction, but generally not a very far-reaching step. The unsupervised study of most books does not assure the effective management of learning. If it did, we might have little need for courses of study, for schools and colleges. Aside from its laboratories, a school might be merely a bookstore. This is not to say that self-guided study of books cannot provide rich opportunities for learning. Many a man has educated himself in this manner; and Thomas Carlyle surely stated a necessary condition, if not a sufficient one, when he said that the true university is a collection of books. But instruction means more than putting such resources at a student's disposal. It means some way to regulate and sequence his activity so that we help him to learn more efficiently than he could do just by diving into the pool of knowledge, as it were, and trying to swim without aid. So we may provide a syllabus, reading assignments, exercises, conferences, examinations, and, of course, the inevitable lectures by the instructor.

But even this paraphernalia often does not provide very efficient conditions for learning. When students listen to a lecture, for example, the instructor is not requiring them to respond in any readily observable way. If he calls for recitation, any one student is responding only a small fraction of the time. And there is no assurance that the other students got the point. Until much too late, the lecturer has no way to know where he has succeeded and where he has failed. We know that students differ greatly in the speed at which they can learn, but a lecture must proceed at a fixed pace. Whatever rate the lecturer adopts in developing new concepts must inevitably be a compromise, aimed perhaps at a hypothetical average listener. This pace is certainly going to be too fast for some and too slow for others. During the usual lecture there are many lapses of attention even on the part of good students. Most students have no way to know until too late whether they are learning what the lecturer intends. These difficulties can be multiplied many times when the instructor is removed from the learner even farther by teaching over radio or television or by a presentation on film.

Programed instruction tries to remedy these deficiencies by going much further in trying to guarantee effective learning than is possible in the usual classroom or mass media lecture. It seeks to manage learning in a more intimate sense and to assure that the student will in fact master the competences which it seeks to create. In effect its approach is to ask how some of the benefits of a private tutor might be made available to all students on an economically feasible basis.

Although the most important concepts underlying programed instruction are quite independent of mechanical devices, its modern development began with a little teaching machine constructed about thirty-five years ago by Professor Sidney Pressey at Ohio State University. With Pressey's machine, the student was taught by having a series of questions put to him, to each of which he had to respond. The machine was designed to be used by one student at a time, so the student could proceed as slowly or as rapidly as his abilities permitted. The questions used by Pressey were cast in multiple-choice form—that is, an incomplete statement followed by several possible answers. The machine had four answer buttons, and the student pressed one of these to indicate which answer he thought was correct for each question. If correct, the machine at once confirmed his answer by going on to the next question. But if he was incorrect, the machine would not go on until the student tried again, and finally selected the correct answer. In the use of one version of Pressey's machine with young children, a piece of candy was automatically dispensed as a reward after a predetermined number of correct responses. In another version of the machine, questions answered correctly once or twice could be dropped out of the series so that the next time through the student's time would not be wasted on material he had already mastered.

Pressey's machines incorporated three basic features found in all subsequent teaching machines as well as in forms of programed instruction that do not employ mechanical devices. These features implement the principles of constant active participation by the student; of immediate correction or confirmation and reward for his successes; and of provision for letting the rate and sometimes the sequence of instruction be determined by the student's own responses.

Sound as these principles appeared to be, Pressey's invention did not attract much attention for nearly three decades. During the late forties and early fifties, however, an independent stream of experimentation developed in the context of research on military training films, using procedures for employing active student response to improve the effectiveness of group and individual instruction. But the spark that touched off the outburst of interest which has mushroomed into a major educational movement was not struck until 1954, when Professor B. F. Skinner of Harvard University published his paper, "The Science of Learning and the Art of Teaching."

Skinner's ideas for a teaching machine were based in large part on his extensive research on reinforcement techniques for training animals in the psychological laboratory. One of the important new concepts he introduced was the idea that any educational subject matter could be analyzed into a large number of very small steps, representing increments of successive approximation to final mastery of the subject. Another concept which followed was that an optimal sequence of steps could be developed and refined on the basis of detailed records of the responses which typical students make to a preliminary version of an instructional program. Together, these concepts embrace much of what we now call programed instruction.

Skinner believed that students should compose their own answers to a self-instructional program rather than choose them from a set of arbitrary alternatives. However, it is difficult to construct a machine that will discriminate the adequacy of the written answers appropriate to more advanced academic work. Accordingly, for such subjects Skinner put the burden of discrimination on the student himself. After the student responds to each item in an instructional series, the correct answer is then presented and the student verifies for himself that the answer he has given is correct. Similar techniques have been employed in student-participation exercises used in the military research on training films. For individual study, Skinner devised a simple teaching machine for presenting printed materials in a way which insured that the student had to give his own response without peeking at the correct answer and merely copying it. Many current teaching machines do no more than this. In any case the teaching is done not by the machine itself but rather by the program it presents to the student.

Step-by-step programed sequences are being developed for building competence in a wide variety of academic subjects and technical skills. These range from elementary skills like spelling and simple arithmetic to highly sophisticated competences requiring judgment to solve complex problems. To do this may require a program consisting of many hundreds or even thousands of steps.

The essential principles on which such programs are based are easy to state, though often difficult to follow: **Decide exactly what you want to teach, then start where the learner is and take him one step at a time to where you want** him to go. Prompt him liberally at first so as to help him come up with appropriate responses, then force him to rely more and more on himself by using what he has learned without prompting. In building gradually toward the desired terminal capability, give him plenty of practice with each new concept or process as it is introduced. Check his progress at each step to correct any errors he may make, and keep him rewarded for each successful new step that he completes.

Now, in themselves there is nothing so very new or startling about these characteristics of programing. To some extent at least, they enter into any good sequence of tutorial instruction, and we can find them advocated in precepts on pedagogy dating back to antiquity. They are suggested, for example, in the Socratic dialogues of Plato and are advocated in *The Great Didactic,* written by Comenius in the seventeenth century. What is new in programed instruction is the degree to which it seeks to implement such principles systematically and reproducibly, and the kind of intensive effort that is being directed to this end.

A bold assumption made by many programers is that whenever the student fails it is not his fault but rather the fault of the program. When a tentative program has been written, it is tried out at once on a few students. The students' errors will show the programer that he has left out an essential element here, presumed too much on the background of the students there. At one point he has taken for granted the understanding of terms that the students haven't previously mastered. At another place he has failed to give sufficient practice or to sharpen a distinction that was clear enough to the programer but that wasn't made explicit enough for the students. All these defects occur in ordinary text and lecture material but they pass undetected because there is not sufficient point-by-point feedback from the student to the instructor. In preparing programed instruction sequences, by contrast, the students' response at each point can tell the programer where he has gone off the track. It is common for a program to go through a number of stages of revision—each stage being based on the detailed data provided by students' responses. Thus in a real sense the students can become co-authors of the program.

Teachers who have tried the exercise of composing a programed-instruction sequence seem to find it a rewarding and revealing experience. Many feel that only by trying to reduce a subject to a programed sequence that will autonomously produce a desired competence does a teacher really learn what the subject matter consists of. Teachers often find out two things in attempting to write a program. First, they find themselves having to define much more precisely than ever before exactly what it is they are trying to teach. Second, they gain a new approach to the development of subject matter skills and a new respect for the difficulties of the student. For it is precisely the students' failures in the early versions of a program that show up the teacher's own errors and that point the way to successful revision.

The attempt to make the program a self-sufficient or automatic teaching instrument has led some to call such sequences auto-instructional programs; their presentation by a mechanical device is sometimes called automated teaching. In using well-programed materials with a suitable teaching machine, several advantages can be realized. The student is kept active and alert. He gets practice in using new terms, concepts, and relationships actively in varied appropriate contexts. If his attention lapses, this is immediately apparent to him. The fact that he is responding frequently in an explicit fashion provides him a constant check on what he is learning. Each time he responds correctly, he is so informed. This can prove to be a very rewarding or reinforcing way to learn. On the other hand, when he makes an error he also knows this immediately, instead of having to wait hours, days, or weeks till the eventual examination. tells him what he hasn't learned. He can work on his own schedule; if he misses a day or a month of school due to illness, he can pick up where he left off when he returns. By the same token, a new resource is provided for adult education and for the child who cannot attend school.

Concern is sometimes expressed over the possibility that programed instruction might result in stereotyped learning and rote performance rather than real understanding of a subject matter. I do not think this fear is well founded. It seems clear by now that programed sequences can lead the student to develop subtle discriminations and generalizations, to apply concepts in new situations, and to utilize them for creative thinking and invention. A related concern is that emphasis on guidance and control over the learner's behavior may develop dependence on being "spoon-fed" by the easy progression of steps in a program. Clearly, the student must eventually learn to cope with an unprogramed world in which he will have to learn from unstructured reading and personal experiences. However, the advantages of "learning by doing" can best be realized only if what the student does is suitably guided during early stages while basic skills and patterns of response are being shaped. This is as true in academic subjects as it is in skills like violin-playing. But any good teaching procedure also must "wean" the student from dependence on the special aids he receives in the early stages of learning. Programed instruction is no exception, and the later stages of a good program must be designed with this end clearly in mind.

The requirement that the student make frequent responses at each step of the way can also serve another function. If we want an auto-instructional sequence that is flexible and adapts most fully to students with varying backgrounds and abilities, we will not want to waste the time of the student in going through all the steps that represent stages of competence he has already mastered. So we may introduce questions which, if answered correctly, will lead to his skipping certain parts of the material. At other points the program sequence may branch off to provide a remedial detour for the benefit of a student whose responses show he has forgotten a previous point. A great deal of experimentation is being done with such variable-sequence or "branching" programs today.

Relatively flexible branching sequences can readily be presented by an arrangement in which each response the student makes is used to modify the program sequence. For printed material, one way of doing this is by a so-called scrambled book. Each page contains a bit of information and a question. As soon as the student answers a question, he turns to another page. But he doesn't go through the pages in order as he would in an ordinary book. Rather, the page to which he is told to turn next depends on what answer he has given to the question on each preceding page. Each incorrect answer leads to a remedial sequence to explain the error before he can go on. Alternatively, a microfilm machine can be used which selects new items automatically in accord with the student's answers. This affords additional control and can keep a record of each student's responses.

Going still further, high-speed digital computers are being employed experimentally to keep track of the students' progress, to store information about their past performance, and to select the sequence that seems most appropriate to each student's individual needs. The choice can be based on a pattern of past responses as well as his current responses to the program. Such electronic devices, time-shared for a number of students, can adapt a sequence on the basis of the promptness as well as correctness of each student's responses. They thus can provide an automated system of programed but adaptive instruction that may approach the flexibility of a skilled human tutor. A related approach is seen in the work of Gordon Pask in applying the concepts of self-organizing systems and game theory as a model for automated teaching devices.

But in many instances we can go a long way toward realizing the advantages of programed instruction without requiring complex instrumentation. The extent to which branching is needed will depend on the extent to which a program can anticipate and head off likely errors, which the student learns to avoid before he gets to the places where they might arise. Carefully designed linear sequences can be used successfully for teaching complex concepts and habits involved in science, mathematics, and certain language skills, with instrumentation no more complex than a specially programed book. Most of the programs currently being written are in this non-branching or linear style. They are presented either in book form or, to obtain better control over the students' behavior, through relatively simple machines which force the student to respond before the correct answer is revealed for comparison. Such machines are currently available at a cost ranging from a few dollars up to fifty dollars or more. Better and cheaper machines are constantly being designed. A word of warning is in order here, however, since even the best of machines is obviously useless without suitable program materials. At the present time, since the field is a new and rapidly developing one, programs are being produced in a variety of formats, not all of which will fit all machines. As in any new field, it will take a while to determine what combinations of program form and machine design are most useful for each instructional purpose.

For very young children who cannot yet read, or in the teaching of oral language skills, use is made of electronic devices for audio presentation. In some cases, also, the student cannot adequately judge the correctness of his own response. Fortunately, in many of the skills we need to teach to young children, the required discriminative responses can be reduced to a few choices—signifying yes–no, larger–smaller, match or non-match, and the like. With ingenuity, such simple discriminative choices can even be used in the teaching of quite sophisticated concepts. In this case, the student need not be required to judge the correctness of his response, because we can use a teaching machine that automatically discriminates the correct from incorrect responses. For programs where the appropriate responses are verbal or numerical, automatic discrimination of responses is also possible if the student can enter his answer on a keyboard instead of writing it by hand. A few such devices are already on the market.

Industrial organizations as well as schools have been quick to see the potentialities of teaching machines and programed instruction for improving the efficiency of their training. A number of extensive self-instruction programs have been written and tried out for employee training. Here the greater specificity of training objectives and accountability of training costs means that savings in efficiency can be more precisely assessed. One careful investigation reports a saving of about 25 per cent in the time required for instruction when programed materials were introduced, as well as a higher level of learning and retention. Some studies in schools report considerably greater gains, others find a less marked advantage.

The amount of improvement found in any study will obviously depend not only on the effectiveness of existing instructional procedures but also on the quality of the programs themselves. There appears to be great variation in the quality of programs now being produced, with quality likely to vary rather directly with the amount of effort expended in developing, trying out, and perfecting the materials. But the very fact that autonomous and reproducible instruction is the goal of a program makes it practical to measure what it teaches, under controlled conditions, to a degree seldom possible with other educational materials. Accordingly, considerable effort is currently being devoted by a joint committee of national educational and psychological associations toward the development of recommended procedures by which program accomplishments can be dependably assessed through the use of standardizable procedures. The inherent advantages and potential for improvability suggest that the limit of attainment for programed instruction is far from having been reached as yet. This inherent improvability has both scientific and technological roots. Aside from the fact that each program is subject to cumulative revision and improvement, the reproducible character of programed instruction makes it a fruitful context for scientific experiments on programing variables. Much research of this kind is being conducted in school systems, universities, and research institutes. From such experiments we can derive principles that will reduce the amount of trial and error in programing. Numerous reports of such experiments are beginning to appear in psychology and education journals.

As progress continues in basic research, in program construction and practical tryout, in assessment procedures and in engineering advances, it seems certain that teachers can be relieved of many routine activities that heretofore have usurped far too much of their time. The liberation of the teacher from the repetitive and often inefficient burdens of teaching, which can now be programed, may well mean that for the first time in the history of mass education, teachers will have sufficient time to work effectively and extensively with individual students in the development of the social and expressive skills that require the teacher's full talents.

I believe the possibilities for further development and application of programed instruction offer both a challenge and a promise for the future. Surely this must apply no less in the improvement of education for developing nations than in the case of nations which are at present technologically more advanced. At the present time the translation of this vision into working actuality is still largely in the experimental stages, and it is accompanied by all the growing pains which characteristically accompany rapid development of new techniques in society. Nevertheless, at the same time that continued research and development of techniques are proceeding in vigorous and sometimes controversy-laden fashion, worthwhile practical applications are already beginning to be witnessed.

SUGGESTIONS FOR FURTHER READING

1 Center for Programed Instruction, Inc. *Programs, '62, A Guide to Programed Instructional Materials Available to Educators by September 1962.* Washington, D.C.: U.S. Government Printing Office, 1962.

2 GREEN, EDWARD J. *The Learning Process and Programed Instruction.* New York: Holt, Rinehart and Winston, 1962.

3 HUGHES, J. L. *Programed Instruction for Schools and Industry.* Chicago: Science Research Associates, 1962.

4 LUMSDAINE, ARTHUR A., and GLASER, ROBERT, eds. *Teaching Machines and Programmed Learning.* Washington, D.C.: Department of Audio-Visual Instruction of the National Education Association, 1960.

5 MARKLE, SUSAN, et al. *A Programed Primer on Programing.* 2 vols. New York: Center for Programed Instruction, Inc., 1961, 1962.

Index

INDEX

Lövaas, Ivar, 121
Lowe, Francis E., 110
Lumsdaine, Arthur A., 43–44, 139–151
Lund, F. L., 48

Maccoby, Eleanor E., 116–127
Maccoby, Nathan, 5, 41–53
McCormick, Thomas C., 110
McGuire, William J., 25, 44, 51
MacKenzie, Robert, 129
McPhee, William N., 99, 129
magazines, radio and, 95, 98
males, persuasibility in, 63
marketing research, 79
Markle, Susan, 151
Mars, "invasion" from, 12
Marsh, C. Paul, 107
Massachusetts Institute of Technology, 3–4, 6
mass audience, "atomized," 80; personalities in, 54–55
mass communication, Army research in, 4–5, 43; crime and violence in, 66, 73–75, 120–121; experiments in, 112–113; as interpersonal, 6; multiple functions and channels in, 105–107; preferences and viewpoints in, 67–68; public taste and, 73–75; reinforcement in, 55, 70, 75; social effects of, 65–76; see also communication; mass media
mass media, channels of, 105–106; children as audience for, 72, 116–127; feedback from, 13; free enterprise and, 69; low intellectual level of, 70; personal influence and, 94–115; and predisposition to change, 71–72; net effect of, 55; research in, 2, 78, 80, 82; in scientific fields, 104; types of, 6–7; voting behavior and, 132–135
mathematics, in communication research, 2
meaning, "common market" in, 37
Measurement of Meaning, The, 28
medical journals, 102–103
Menzel, Herbert, 3, 94–115
Merton, Robert K., 115
message, distortion of, 78; frame of reference and, 6–9; transmission process and, 10–11

Meyersohn, Rolf B., 100
Michigan, University of, 129
Michigan State University, 6
Mill, John Stuart, 41
mobility, of opinion leaders, 101
motivation, behavior and, 18
movies, aggressive behavior and, 122–123; documentary, 54; magazine reviews of, 99; as mass media, 67
"Mr. Biggott" studies, 11
music lovers, 72
Mussen, Paul, 122

narcissistic character, 60
need satisfaction, 49
new ideas and practices, diffusion of, 77–91
newspapers, as mass media, 6, 69; voting behavior and, 134–136
Nixon, Richard M., 137
Nuffield TV studies, 117–118

Ohio State University, 141
opinion, action and, 19; climates of, 107–109; dissonance reduction and, 26; public, 78–79
opinion change, predisposition to, 58–59, 69, 71–72
opinion leaders, characteristics of, 96–100; personal influence of, 99; physicians as, 84; radio and magazines as aid to, 97–98; role of, 109–112; in two-step communication process, 96
oral vaccine, adoption of, 90
Order of Presentation in Persuasion, The, 43, 48
Osgood, Charles E., 28–40
other-directed personalities, 62

panel method of study, 80, 96, 129
paranoid tendencies, 60
Pask, Gordon, 147
passive-dependent personality, 62
peace, persuasibility and, 61
perceptions, mapping of, 52
personal influence, 112; voting and, 134; see also "influentials"; personality
personality, persuasibility and, 47, 54–64
Personality and Persuasibility, 47

156

signs and symbols, 7; connotative
meaning of, 38
Skinner, B. F., 141–142
small-group research, 82
Smith, Kate, bond sales of, 12
smoking, 68
"snowball" sampling, 81–83
social anthropology, 78
social change, 77
social effects, of mass communication,
65–76
social institutions, and predisposition
to change, 75
social integration, change and, 86–88
Social Judgment, 43, 52
social pressure, 62
social psychologists, 55; diffusion
and, 78
social science disciplines, 77–78
social status, choice and, 82
society, communication and, 13–14
sociology, diffusion and, 78; rural, 85
sociometric questionnaire, 83–84
Socrates, 42
sophistry, meaning of, 41
Soviet Union, oral vaccine in, 90;
war origin study and, 43
Stanford University, 6
structural theory, Rosenberg's, 50–51

taste, levels of, 66, 74
teaching machines, 139–150; understanding and, 146
television, aggressive behavior and,
120–122; childhood experience and,
126–127; crime and violence in,
120–121; educational programs in,
123; effect of on children, 116–127;
as escape from tension, 126; as mass
communication, 6; in 1960 presidential campaign, 111, 136–137;
personality of audience in, 54;
radio listening and, 120; repetitive

themes in, 124; and unit of adoption, 89–90
television programs, selective exposure to, 71
theory, vs. practice, 5
Theory of Cognitive Dissonance, A,
25
Thomson, R. V., 118
Thurstone, L. L., 123–124
timidity, 61
toothache, fear arousal and, 45–46
tribe, communications in, 14
two-party system, 131

United States, communications research in, 1–15; *see also* presidential election; voting behavior

value patterns, change and, 89
verbal opposites, 29
voting behavior, 65–67, 81, 96, 99;
communication and, 128–137; decision-making in, 131–132; in 1960
campaign, 111–112; research in, 80–
81

war origins, study of, 43
Weiss, Walter, 45
Welles, Orson, 12
Western programs, TV, 118
Whorf, Benjamin Lee, 34
Wilkening, Eugene A., 107
withdrawal, 60
word-of-mouth communication, 78
103
world peace, empathy as factor in, 61
World War II, communication re-
search in, 4, 43; primary groups
and, 109
Wundt, Wilhelm, 37

Yale University, Communication and
Attitude Change Program at, 4–6,
43, 55